More Pr
ALIEN O

T0260771

"*Alien Oceans* offers a historical look—as well as a peek into the future—at one of the most exciting aspects of space exploration."
—SID PERKINS, *Science News*

"[Hand] describes why studying Earth's own ocean is a crucial chapter in the quest to explore the shores of extraterrestrial seas."
—NADIA DRAKE, *National Geographic*

"What is so captivating about this book is that it isn't just a solid survey of what we've learned in recent decades about the icy moons, but that the narrative is told by an active researcher deeply embedded in these endeavours. Through Hand's eyes we meet many of the key personalities involved and feel the sting of disappointment at cancelled funding or a malfunctioning probe, as well as the soaring excitement of a new discovery."
—LEWIS DARTNELL, *BBC Sky at Night*

"This book would make anyone excited about space."
—RACHEL DEHNING, *Manhattan Book Review*

"Fascinating. . . . Hand has written a thoughtful and thought-provoking treatise on the many facets that are being pursued in our quest to discover new worlds and search for life beyond our atmosphere."
—MILBRY C. POLK, *Explorers Journal*

"Engaging. . . . To paraphrase Hamlet, there are more things in heaven and Earth than are dreamt in our philosophy. Hand calls on us to probe the depths of alien oceans to discover them."
—ROBERT ZUBRIN, *National Review*

"In this delightful book, Kevin Peter Hand takes readers from the depths of Earth's oceans to those of the outer solar system, describing encounters with magical, alien-like creatures at the bottom of the Atlantic and offering informed speculations about what life could be like in the subsurface oceans of faraway moons. Recounting the story of how we discovered these alien oceans, he gives us a peek at the lives and personalities of some of the scientists who pieced together all the clues. His explanations are full of engaging analogies that will help general readers understand the science needed to think rigorously about life as we know it—and as we do not yet know it."
—JILL TARTER, SETI Institute

"Kevin Peter Hand is an explorer—an explorer of the arctic, of Earth's deep oceans, and of outer space. Now he longs to explore the vast oceans beneath the ice of moons of Jupiter and Saturn. In his book *Alien Oceans*, he straps you in and takes you along for the ride."
—CHRISTOPHER F. CHYBA, Princeton University

"Hand humanizes the science behind the search for life on icy worlds in our solar system and beyond."
—GORDON SOUTHAM, University of Queensland

"Hand provides general readers with the tools for understanding the search for life on ocean worlds beyond Earth. His conclusion is clear: go explore these exciting worlds, for they may hold the secrets to the preponderance or scarcity of life in the universe and the origins of life on our own planet."
—J. HUNTER WAITE, Southwest Research Institute

ALIEN OCEANS

ALIEN OCEANS

THE SEARCH
FOR LIFE
IN THE DEPTHS
OF SPACE

KEVIN PETER HAND

PRINCETON UNIVERSITY PRESS
PRINCETON AND OXFORD

Copyright © 2020 by Princeton University Press

Requests for permission to reproduce material from this work
should be sent to permissions@press.princeton.edu

Published by Princeton University Press
41 William Street, Princeton, New Jersey 08540
6 Oxford Street, Woodstock, Oxfordshire OX20 1TR

press.princeton.edu

First paperback printing, 2021
Paper ISBN 978-0-691-22728-3
Cloth ISBN 978-0-691-17951-3
ISBN (e-book) 978-0-691-18964-2

British Library Cataloging-in-Publication Data is available

Editorial: Jessica Yao and Arthur Werneck
Production Editorial: Brigitte Pelner
Jacket/Cover Design by Amanda Weiss
Production: Jacqueline Poirier
Publicity: Sara Henning-Stout (US) and Katie Lewis (UK)
Copyeditor: Susan Matheson

This book has been composed in Arno

Printed in the United States of America

TO MY PARENTS,

Peter and Marybeth Hand,

for providing a childhood

few in fences but full of nets

CONTENTS

Acknowledgments   xi

PART I: OCEANS NEAR AND FAR

Prologue: The Bottom

3

CHAPTER 1
Ocean Worlds on Earth and Beyond

12

CHAPTER 2
The New Goldilocks

25

PART II: DISCOVERING AN OCEAN IN THREE EASY PIECES

CHAPTER 3
The Rainbow Connection

49

CHAPTER 4
Babysitting a Spacecraft

69

CHAPTER 5
How I Learned to Love Airport Security

79

CHAPTER 6
Lady with a Veil

96

CHAPTER 7

The Queen of Carbon

108

CHAPTER 8

Oceans Everywhere

120

PART III: THE JOURNEY FROM HABITABLE TO INHABITED

CHAPTER 9

Becoming Inhabited

137

CHAPTER 10

Origins in an Alien Ocean

151

CHAPTER 11

Building an Ocean World Biosphere

160

CHAPTER 12

The Octopus and the Hammer

184

CHAPTER 13

A Periodic Table for Life

207

PART IV: THE NEXT STEPS

CHAPTER 14

Seeking Signs of Life

229

CHAPTER 15
A New Age of Ocean Exploration
245

Endnotes   261
Index   271

ACKNOWLEDGMENTS

This book slowly bloomed from my brain over the course of many, many years. I am indebted to all who helped nurture my brain, culture my curiosity, and encourage my desire to explore. Foremost are my parents, Peter and Marybeth Hand, and my siblings, Sean, Meaghan, and PJ. The support of my extended family, a.k.a. the Thanksgiving crew, has also been very important throughout the years, especially the editorial help of my aunt, Kathy Barry, whose eagle eye for grammar was incredibly helpful in early drafts.

I credit growing up in the small town of Manchester, Vermont, with its clear night skies and harsh but forgiving Green Mountains, to my lifelong obsession with the stars and the search for life beyond Earth. Several teachers from my early years deserve credit for blending science, math, art, and writing together in a way that still serves as the basis for how I think and communicate: Mike Ceperan, Anita Dorfman, Marilyn Hopkins, Jean Noble, Margaret Connor, Donna Williams, Lucille Jordan, Betsy Hubner, and Bob and Bev Leslie. Thanks also to Trey Spencer and Louie Dean for providing many useful lessons on and off the mountain.

Deep gratitude to Michael Ferraro, Tyler Stableford, Steve Drunsic, Alicia Williams, Tom Nordheim, Dan Berisford, Tom Painter, Charles Elachi, Dan McCleese, Firous Naderi, Sarah Horst, Sam and Bettyrae Eisenstein, Tony Krantz, Sean Carroll, and Jennifer Ouellette for various degrees of moral support, editorial comments, and guidance along the way. Thanks also to the crew at the Colorado Wine Co., Café de Leche, and the Hermosillo; many glasses of wine, cups of coffee, and pints of beer were consumed during the production of this manuscript.

I am grateful for the hard work of my colleagues at Princeton University Press, in particular the editorial brilliance of Jessica Yao, as well as that of Brigitte Pelner, Arthur Werneck, and, early on, Eric Henney.

Many friends and colleagues at the Jet Propulsion Lab have inspired me along the way, most of all the entire Europa Lander mission concept team and my crew in the Ocean Worlds Lab. I have been very fortunate to work with such inspiring and talented people.

To Margaret Kivelson, John Casani, and Bob Carlson I owe thanks for your time for various conversations, interviews, and stories. To Bob Carlson I owe an additional thanks, as he took a chance on me many years ago, and that has led to years of great work together, as well as numerous adventures.

A special thanks to James Cameron for immersing me into the world of deep ocean exploration. Without his vision, a scientist like me never would have had the chance to dive down to the hydrothermal vents.

Lastly, endless appreciation goes to my former advisor and longtime friend, Chris Chyba. I was very fortunate to have the guidance of such a gifted and gracious human being early in my career. Science and exploration march forward in myriad ways, with unpredictable ups and downs, and sometimes with resistance in all directions. Finding kindred spirits with whom to share the big picture makes it all worthwhile.

Thanks to everyone above, and all in my life, who have helped keep my curiosity and sense of wonder for the cosmos pure. I hope this book captures and conveys some of that wonder.

PART I

OCEANS NEAR AND FAR

THE BOTTOM

We were stuck on the bottom. Batteries were running low. Our air was running out. We had no way to communicate to the other submersible or to the team on the boat some 10,000 feet above us. We were nestled in the metal sphere of our tiny submersible, perched on some rocks at the bottom of the Atlantic Ocean.

This was my first trip to the ocean floor, and it had the makings to be my last.

Yet somehow it seemed peaceful. With what little light we had left I was able to look out through the porthole of three-inch glass and see a long, red creature exploring the surface of a rock, perhaps looking for its next meal. There it was, going about its business, with no concern or awareness of our precarious situation.

It's easy for things to turn surreal and serene in a tiny submarine. Our brains have no way of processing the reality of the situation: several thousand pounds of pressure per square inch, a landscape revealed only by the limited lights on the sub, odd sounds and whirls of the machine that's protecting you from a gruesome, watery death. Unlike experiencing a fear of heights or a bone-dry desert, there's nothing in our *Homo sapiens* software that knows what to do when cooped up in a metal ball at the bottom of the ocean. We had no connection to a mouthpiece and breathing regulator like scuba divers and had no need for decompression on the way up. In a submersible you simply breathe the air around you. Sure, there's the potential for claustrophobia with three people crammed into a sphere two meters in diameter and the darkness of a world cut off from sunlight; but if you can get beyond that, it's really quite nice. That is, if everything is working as it should.

I actually had no business being down there. First, as a human I was clearly out of my biological comfort zone; I needed technology to make this trip possible. But second, and perhaps more important, my professional realm had long been that of the stars, planets, and moons. A childhood obsession with aliens had led me down a path to studying Europa, an ice-covered moon of Jupiter that had recently been revealed to harbor a vast subsurface ocean. I was in the midst of my PhD studying Europa's physics and chemistry when fellow space nerd and longtime friend George Whitesides rang me to gauge my interest in an exciting project: James Cameron, filmmaker of *Titanic*, *Terminator*, and many other successful movies, was looking for a young scientist to talk about Europa while exploring the depths of our ocean. Would I be interested in potentially joining the expedition? It was not your everyday phone call.

The year was 2003 and Cameron wanted to make a film about the deep sea and the prospect of searching for life in Europa's ocean. The team would explore the seafloors of the Atlantic and Pacific Oceans, studying how life survives in the dark depths—conditions that might be comparable to those found on Europa. My role would be to help connect ocean exploration with the search for life beyond Earth. The deep-sea hydrothermal vents that we would explore serve as chemical oases for life in the ocean's depths and provide some guidance in our search for habitable environments beyond Earth.

At the time, given where I was in my studies, it was a difficult decision to make. My primary research would have to take a backseat while I was out at sea. Did I want to add this diversion during a time when I was very focused on my PhD? The depths of our ocean are potentially a good analog for what might be going on within Europa's ocean, so from that standpoint I could at least justify the time away.

In retrospect, it should have been a simple decision. If you ever get the chance to go to the bottom of the ocean, take it. Don't think twice; just say yes. Your brain will be changed forever.

My own internal deliberations were clarified by the sage words of mentor, musician, and microbiologist extraordinaire, Professor Ken Nealson of the University of Southern California. Standing on a dock one evening on Catalina Island, Ken grabbed my shoulder and said, "You're overthinking it! If there's even the slightest chance you get to go to the bottom of the ocean, you better go!"

And so, a month after that phone call from George, I found myself on the Russian research vessel *Keldysh*, floating above the middle of the Atlantic Ocean, preparing to explore the darkness below instead of the stars above.

The ocean had long been magical to me—not just because of its vastness and great depths but because it was a place with which I was largely unfamiliar. I had grown up in the landlocked state of Vermont. Set me down in the mountains or in a cave and I'd do fine, but the ocean was a foreign environment to me. I wasn't afraid, per se—there was no *Jaws* music running on a loop in my head. Rather it was that lack of familiarity: I did not know how to read the clouds and winds, and there would be no trees whispering about the weather, no mountains to break the sky. Only an arc of the horizon, with water from end to end, the world below hidden by the waves above.

I was fascinated. Cruising from the port in Halifax to the middle of the Atlantic Ocean, I spent many hours—day and night—sitting on the bow of the *Keldysh*, gazing out at the water, trying to build an innate sense of my new environment. But it was alien. The vast expanse of the ocean was well beyond my comfort zone—exciting and intimidating at the same time.

With a combined sense of uncertainty and anticipation, I turned my attention to the machines that would take us into this extreme environment—the Russian *Mir* submersibles (in Russian, *mir* means peace or world). I spent time in the machine shop on the *Keldysh*, communicating with hand gestures and the occasional Russian word. I was eager to get a sense of how the submersibles work—and what could go wrong. I had no idea what to look for in a good sub, but the engineers were very patient. I soon learned that Russian engineers are nothing if not robust and thorough: there was always a backup plan, and a backup to the backup plan, and so on.

The basic goal of the physics of a submersible is pretty straightforward: (1) don't get crushed, and (2) make sure you can float when you need to. Unlike space exploration, gravity is your friend when exploring the ocean floor. To get back and forth from space requires any number of variations on rocket engines, heat shields, parachutes, and wings. To get back and forth from the seafloor, the general idea is to carry a weight on the way down and then drop it when you want to come up. Although the change in pressure in the ocean is fairly extreme, there's not a huge temperature difference, and you're never traveling that fast. As long as your sub can withstand the pressure and is buoyant once you drop the weights, you'll rise to the top of the ocean like a cork.

The basic underlying simplicity of how a submersible moves up and down through the water was central to my getting comfortable with the *Mir* subs. Genya, Viktor, and Anatoli—three of the pilots and engineers who endured my questioning—explained the various backup systems and redundancies the subs boasted. For the most part, the subs followed the prime KISS rule of engineering design: Keep It Simple, Stupid. They had relatively few moving parts, and the electronics seemed like hardy relics from the Cold War. Nevertheless, my stream of "What ifs?" eventually led to the worst-case scenario: What if you're many kilometers down at the bottom of the ocean and your power fails, your thrusters fail, your communications link fails, and you start running out of air? You're just sitting in a fancy hunk of metal trapped at the bottom of the ocean. What then?

Not surprisingly, there was a plan. In that scenario, you lift up one of the seats in the sub and find a big wrench. That wrench is used to loosen a large nut on a bolt that is connected to a weight. Once that weight has dropped, the sub becomes positively buoyant. It should start to rise off the bottom of the ocean and gradually accelerate as it rises upward. According to the engineers, by the time the sub reaches the surface, it will have amassed so much momentum that it would likely pop out of the ocean and into the air, rising a few meters above the ocean surface. It's not pretty or high-tech, but at least you're not dead on the bottom.

All of this was going through my head as we sat stranded on the sea-floor in our dying submersible. I was starting to think more and more about that wrench. Was I certain it was under the seat? I should have double-checked before we left.

Up until that point, everything had gone quite well. After reaching our location near the middle of the Atlantic Ocean, we had dived from the ship in two subs—Cameron and two team members in one, me and two team members in the other. Our goal was to explore hydrothermal vents on the side of a seamount called Menez Gwen (basically a volcano on the bottom of the ocean).

Being in a submersible at the bottom of the ocean feels like a hot air balloon ride, scuba dive, and space mission rolled into one. (Mind you, of the three, I've only ever scuba dived.) Motions are, for the most part, slow and smooth. The Russian *Mir* submersibles offer a small (approximately eight-inch) porthole from which to peer out into the depths: one porthole for each occupant, three in total. In the center is the pilot and to the left and right are two passengers. You can't stand up, but there's enough room for a cramped, hunched-over shuffle if one person needs to trade places with another. Instruments, buttons, and switches along the interior are reminiscent of the 1970s and 1980s, when these subs were built and first deployed. Seats with blue vinyl cushions form a tiny, communal U-shape, but there's not enough space for all three people to have their legs on the floor at the same time. Usually, the pilot sits upright in the middle, hands on the control sticks, while the two passengers lie on their sides, which they have to do to peer out the windows.

On the way down to the bottom, when the machine is dropping like a stone, you pass rapidly through the photic zone—the uppermost layer of water, approximately 300 meters deep, through which sunlight shines and in which life thrives off of photosynthetic organisms like phytoplankton.

As you descend through this region, light begins to fade. Blue goes to black. The sub starts to cool. You can't feel that you're falling, but the sounds from the acoustic communication system serve as a metronome reminding you of the distance between you and the rest of the world. Loud pings bounce between the sub and the *Keldysh* every few seconds—exactly the clichéd pinging you would expect to hear while sitting in a submersible but a little higher in pitch and shorter in duration. Every so often, words trickle through the speaker: a sentence or two of Russian, significantly broken up by its journey through the water. For me, with my limited Russian vocabulary, the foreign patter and pings made the environment feel even more alien.

As we sank, I pressed my face up against the porthole and wrapped my head in a towel, trying to prevent light from inside the sub from contaminating my view outside. This part was familiar to me—astronomers gazing at the night sky are always protective of their night vision. The better adapted you are to the dark, the more stars you can see in the night sky.

Here, however, I was scouring the depths for a glimpse of bioluminescence from the host of bizarre creatures that populate our ocean's depths and that give off pulses of light as they, or things around them, move. The plummeting submersible created a shock wave of bioluminescence radiating away from us. Creatures large and small, from jellyfish to microbes, flashed. It was a sight I'll never forget, and it's one that I treasured on each of my nine dives. I got into the habit of donning earphones, pairing the biological fireworks with the music of Radiohead and Pink Floyd. I half-expected to find those bands playing on the ocean floor.

So far, this dive, my first, had been very successful. After some initial exploration of the seafloor, we managed to track down a site rich with hydrothermal vents. For hours we cruised over the flanks of Menez Gwen and collected samples with the robotic arm from the gushing superheated vents. Microbes, mussels, and fish feasted on the bounty of these chemical oases.

Finding this spot on the bottom of the ocean may not seem like a big achievement, but it really was. All of those mapping luxuries that we employ on the surface of the Earth are useless in the ocean. The wavelengths we use for GPS navigation barely penetrate beyond a few millimeters into the ocean's surface. In fact, the ocean is a bad place for transmitting just about any wavelength in the electromagnetic spectrum. Water—the key ingredient to life as we know it—turns out to be very good at hiding much of our own planet from us. Liquid water readily absorbs light, from short to long wavelengths, and thus we can't "see" or communicate from the bottom of the ocean. This simple fact has confounded engineers for decades: no electronic mode of navigation or communication works underwater—no cell phones, no Wi-Fi, no GPS, no AM, no FM, no ham (amateur) radio, nothing.

Nothing makes it through the ocean over very long distances, except sound. This is partly why whales and dolphins use sound to communicate. And it's why those acoustic pings kept coming through into our sub from the team on the *Keldysh*, checking to see whether we were still there and whether everything was okay.

But it wasn't okay. We had overextended our stay. It had been a long dive and we had explored quite a large area of the seamount. Both *Mir 1* and *Mir 2* were at the bottom, and we had worked in tandem to sample and film much of the region. But now our sub (*Mir 2*) had lost track of *Mir 1*, and we were starting to run low on battery power. Viktor, the pilot, decided it was not a good idea to move or "swim" around anymore since the thrusters used precious battery power.

In addition, our communications link was failing, and our low batteries meant that we could not send strong signals back to the *Keldysh* or to *Mir 1*. On top of everything else, the carbon dioxide scrubber in our submersible was starting to fail, turning our little sphere into a ball of toxic gas.

So there we sat, on the bottom of the ocean, troubleshooting our situation. Although it was a bit scary, I took solace in the wrench. If nothing else, we would pop up out of the ocean and eventually the *Keldysh* crew would find us. Trouble was, depending on the time of day and state of the ocean waves, it might take many hours for the *Keldysh* to find us.

Viktor tried every trick in the book before resigning himself to relaxing in the sub. From his demeanor, either he had accepted the fate of a watery death for us all or he knew that some sort of search protocol would kick in and our job was to sit and wait. My mind continued to wander, as I looked out the porthole, staring at the red shrimp-like creature. My imagination drifted from our ocean to the possibility of oceans beyond Earth.

The scene through that porthole could be what "home" looks like to most of life in our Universe. Deep, dark, seemingly desolate ocean floors may be some of the best real estate for biology. Recent explorations of our solar system have taught us that—while planets like Earth may be comparatively rare (one per solar system, if you're lucky)—worlds with deep oceans, covered with ice and cut off from any sky or atmosphere above, could be ubiquitous.

In our solar system, these worlds are actually moons of the giant planets, with names like Europa, Ganymede, Callisto, Titan, Enceladus, and Triton. These are worlds that likely harbor oceans of liquid water today, right now, and their oceans have likely been in existence for much of the history of the solar system. A few of these worlds—Europa, Enceladus, and Titan—could even be hospitable to life as we know it.

These ice-covered oceans have no beaches or sandy shores, but they are potentially wonderful, and plentiful, places to call home. The dark depths of those distant oceans may look similar to the deepest regions of our own ocean. Microbes and sea creatures that inhabit our ocean depths might do fine under the physical and chemical conditions thought to exist within the oceans of Europa, Enceladus, and Titan. Throughout this book, we explore this link between the extremes of life on Earth and the physics and chemistry that may make for habitable ocean worlds elsewhere. We examine in detail how we think we know that oceans exist beyond Earth, beneath the icy shells of moons in the outer solar system. And finally, we consider what that means for discovering a separate, independent origin of life beyond Earth.

But before that journey begins, I need to get off the bottom of the ocean.

Although it seemed as though we were stranded on the ocean floor for a long time, it could not have been more than 45 minutes. It had grown humid inside the sub from the condensed water of our own biology. Each breath we took converted precious oxygen into water vapor and carbon dioxide. The curved orange walls of the sub were dripping. It was cold, damp, dark, and quiet. Meanwhile, that little red creature continued to live its life just outside our porthole—blithely unaware of the peril we faced and perfectly happy to scramble across the rocks.

Then, through the porthole a faint light appeared. The light grew, slowly meandering on a path that brought it closer and closer to us.

Finally, the light revealed its source—our sister sub, the *Mir 1*. They had found us.

Knowing that Cameron and the *Mir 1* team were running cameras that would be able to see us, we wrote a note on a piece of paper and pressed it against a porthole. The note said simply: "Must surface." We watched the camera pan and tilt, inspecting us, assessing the situation. It seemed like they had gotten the message. After a few more attempts at communication with paper and with the acoustic system, Viktor felt confident that the *Keldysh* would be ready for us, and we could now lift off from the bottom and ascend.

The trip back to the surface has a bizarre elegance. Unlike spaceflight, where the return to Earth involves an intense and fiery trip through the atmosphere, followed by aggressive thrusters to fight gravity's attempt to turn you into ash, the journey back from the bottom of the ocean feels like a slow elevator ride. The laws of physics are a gentle friend, not a foe, as the sub's buoyancy guides you back home. Gravity does all the work— no thrusters needed, no engines fired, no fear of a parachute not deploying.

If you happen to ascend during daytime, the dark abyss eventually gives way to subtle hints of a star above. Black fades to blue as the most intrepid of the Sun's rays pierce through the water. With each moment, blue pushes black farther down, and an ocean fed by sunlight emerges. In the final moment, the sub rises and then falls as it bobs up onto the surface of the ocean, once more touching the atmosphere. Sunlight blasts through the portholes, reminding your brain of where it really belongs.

Now safely returned to the surface, we sat bobbing in a tiny speck of orange and white on the vast blue of the ocean. A tiny but considerable speck, returned from an otherworldly experience in the depths, waiting to be picked up by the *Keldysh* and brought on deck by its massive crane. One journey, to one spot, was complete—but with so much more to explore and discover.

CHAPTER 1

OCEAN WORLDS ON EARTH AND BEYOND

If we have learned anything from life on Earth, it is that where you find liquid water, you generally find life. Water is essential to all life as we know it. It is the solvent, the watery broth that makes possible all the chemistry in our cells. Water dissolves many of the compounds that life, large and small, needs to grow and metabolize. Every living cell is a tiny bag of water in which the complex operations of life take place. Thus, as we search for life elsewhere in the solar system, we are primarily searching for places where liquid water can be found today or where it might have existed in the past.

The story of the search for life beyond Earth is, in part, the story of our planet, the pale blue dot,[1] reaching out into space, seeking signs of life on other worlds. Like a plant stretching vines out into its environment, our little planet has been sending its robotic emissaries out in spiral tendrils that circle other planets, probing for answers and sending back information.

We humans have been exploring our solar system with robotic vehicles for over 55 years. The first robotic mission to another planet was the flyby of Venus by the *Mariner 2* spacecraft on December 14, 1962. Since then, we have sent an armada of spacecraft to study the Sun and a variety of planets, moons, asteroids, and comets, most of which are in the inner reaches of our solar system. Over that same period, we have sent only eight spacecraft beyond the asteroid belt to study the many worlds in the outer reaches of the solar system.

Spacecraft that have gone beyond the asteroid belt—*Pioneer, Voyager, Galileo, Cassini, New Horizons, and Juno*—have revealed something profound about what it means for a world to be habitable. The data returned from those missions have served to revolutionize our understanding of where liquid water exists in our solar system, and by extension, where life might find a home.

We now have good reason to predict that at least six moons of the outer solar system likely harbor liquid water oceans beneath their icy crusts. These are oceans that exist today, and in several cases we have good reason to predict that they have been in existence for much of the history of the solar system. Three of these ocean worlds—Europa, Ganymede, and Callisto—orbit Jupiter. They are three of the four large moons first discovered in 1610 by Galileo. The fourth moon, Io, is the most volcanically active body in the solar system and does not have water. At least two more ocean worlds, Titan and Enceladus, orbit Saturn. Neptune's curious moon Triton, with an orbit opposite to the direction it rotates, also shows hints of an ocean below.

These are only the worlds for which we've been able to collect considerable data and evidence. Many more worlds could well harbor oceans. Pluto may hide a liquid mixture of water, ammonia, and methane, creating a bizarrely cold ocean of truly alien chemistry. The odd assembly of moons around Uranus—such as Ariel and Miranda—might also have subsurface oceans.

Finally, throughout the history of the solar system, ocean worlds may have come and gone; for example, the large asteroid Ceres likely had a liquid water ocean for much of its early history. Mars and Venus may also have had oceans previously. Early in our solar system's history, oceans might have been commonplace, be they on the surface of worlds like Venus, Earth, and Mars, or deep beneath icy crusts of worlds in the asteroid belt and beyond. Today, however, it is the outer solar system that harbors the most liquid water.

This distinction—between liquid water *in the past* and liquid water *in the present*—is important. If we really want to understand what makes any alien organism tick, then we need to find life that is alive today and that requires the presence and persistence of liquid water.

The molecules of life (e.g., DNA and RNA) don't last long in the rock record; they break down within thousands to millions of years, which, geologically speaking, is a short amount of time. Bones and other mineral structures of life can stick around much longer and form fossils. Fossils are great, but they only tell you so much about the organisms from which they formed.

As an example, Mars may have been very habitable roughly 3.5 billion years ago. Robotic vehicles, including the rovers *Spirit*, *Opportunity*, and *Curiosity*, have revealed that chemically rich lakes, and possibly vast oceans, populated the Martian surface. If life did arise within those liquid water environments, then some chemical or structural "fossils" might remain preserved within rocks from those ancient times. We would not, however, be able to extract any large molecules like DNA from those fossils. Our search for life on Mars is largely focused on scouring the rocks for any signs of ancient life that went extinct long ago.

Make no mistake, if we were to find rocks on Mars that showed signs of ancient life, it would be an extraordinary discovery. However, I would be left wanting more. I want to find life that is alive today, life that is *extant* as opposed to *extinct*.

This is important because I really want to understand *how life works*. What is the biochemistry that drives life on another world? On Earth, everything runs on DNA, RNA, ATP, and proteins. Darwinian evolution through natural selection has led to our amazing biosphere. The same fundamental biochemistry connects all of life's wild diversity. From the most extreme microbe to the craziest rock-and-roll star, we all have the DNA, RNA, ATP, and protein paradigm at our root. I want to know if there could be another way.

Can life work with some other fundamental biochemistry? Is it easy or hard for life to begin? Does the biochemistry of the origin of life converge toward DNA and RNA? Or were there contingencies that made these the best molecules for life on Earth but perhaps not on other worlds? If we were to find extant life in an ocean world, we could begin to truly answer these questions.

At an even higher level, consider the big picture of human knowledge.

When Galileo first turned his telescope toward the night sky and began charting the faint points of light he saw around Jupiter, he set in motion a revolution in physics. Night after night he drew Jupiter and the arrangement of these points of light. At first, he concluded that they must be stars that he could not see with the naked eye. He even named them the "stars of Medici" in honor of the Medici family since they were funding his research (Galileo was no idiot).

But through his diligent charting of these points of light, Galileo soon realized that they were not stars; they were moons orbiting Jupiter. His discovery got him into deep trouble with the Spanish Inquisition, and he ended up under house arrest. The idea that a celestial body would orbit anything other than the Earth was heretical. The world view at the time was framed around Aristotelian cosmology—the Earth is at the center of the universe and everything revolves around the Earth. Galileo's discovery put him at odds with this world view and provided strong evidence for the growing Copernican Revolution, the idea that the planets orbit the Sun and that the stars we see could well be suns with planets of their own.

In the decades that followed Galileo, advances in math and physics would lead to an appreciation that the laws of physics *work* beyond Earth. Gravity, energy, and momentum govern objects here on Earth as well as on worlds and wonders beyond.

In the century that followed these developments, the field of chemistry would grow and expand, eventually yielding instruments that could tell us the composition of the Sun, stars, and planets. The elements of the periodic table made up everything on Earth and beyond. Chemistry, too, works beyond Earth.

In the twentieth century, with the advent of the space age, our human and robotic explorers to the Moon, Venus, Mars, Mercury, and a host of asteroids would reveal that the principles of geology work beyond Earth. Rocks, minerals, mountains, and volcanoes populate our solar system and beyond.

But when it comes to biology, we have yet to make that leap. *Does biology work beyond Earth?* Does the phenomenon we know and love and call life *work* beyond Earth? It is the phenomenon that defines us, and yet we do not know whether it is a universal phenomenon. It is a simple but central question that lies at the heart of who we are, where we come from, and what kind of universe we live in.

Is biology an incredibly rare phenomenon, or does life arise wherever the conditions are right? Do we live in a biological universe?

We don't yet know. But for the first time in the history of humanity, we can do this great experiment. We have the tools and technology to explore and see whether life has taken hold within the distant oceans of our solar system.

SEARCHING FOR A SECOND ORIGIN

To answer these questions, we need to explore places where life could be alive today, and where the ingredients for life have had enough time to catalyze a second, independent origin of life.

This aspect of a second, independent origin is key. Take Mars again. Even if we were to find signs of life on Mars, there are limits to what we'd be able to conclude about that life form, and about life more generally. Mars and Earth are simply too close and too friendly, trading rocks since early childhood. When the solar system and planets were relatively young, large asteroids and comets bombarded Earth and Mars with regularity, scooping out craters and spraying ejecta into space. Some of this debris would have escaped Earth's gravity and may have ended up on a trajectory that eventually impacted Mars (or vice versa).

We know that life was abundant on Earth during many of these impact events, and thus it is not unreasonable to expect that some of the ejecta were vehicles for microbial hitchhikers—some few of which could (with a small probability) have survived the trip through space and the impact on Mars. Even if just a few microbes per rock survived, there were enough impacts and enough ejecta that the total number of Earth microbes delivered to Mars has been calculated to be in the range of tens of billions of cells over the history of the solar system. If one of those Earth rocks came careening through the Martian atmosphere about 3 billion years ago, it could have dropped into an ocean or lake on Mars, and any surviving microbes on board might have found themselves a nice new home on the red planet.

This possibility, however remote, would make it difficult to be completely confident that any life we discovered on Mars arose separately and independently—in other words, that it was really Martian. Life on Mars could be from Earth, and vice versa.

If we found fossils of microbes in ancient rocks on Mars, we would not be able to determine if that life used DNA or some other biochemistry. Lacking strong evidence for the more extraordinary claim of a second, independent origin of life on Mars, we would potentially have to conclude that Martian life came from Earth.

Indeed, even if we found extant life in the Martian surface or subsurface, there would still be significant potential for confusion as to where that life came from. Imagine that we found living microbes in the Martian permafrost or in some deep aquifer, and imagine even further that we discovered that those organisms also used DNA-based biochemistry. Even if we were unable to connect it to our tree of life, this shared biochemistry would force us to consider that Earth life and Martian life may have shared a common origin, whether life was transported from Earth to Mars or the other way around.

Although it's possible that such DNA-based life on Mars arose through convergent biochemical evolution, it would be hard to differentiate that scenario, and ultimately we would still not have conclusive evidence for a second origin. The only truly robust support for a second origin of life on Mars would be the discovery of extant life with a non-DNA-based biochemistry. Even then, there would still be a few scenarios to consider, and discard, that could implicate the Earth as the place of origin.

The ocean worlds of the outer solar system do not suffer these pitfalls. First, by focusing on worlds with liquid water oceans, we are focusing on worlds that could harbor extant life; and thus we could study their biochemistry in detail. Second, the "seeding problem" is almost negligible. Very few rocks ejected from the Earth could make it all the way to Jupiter and Saturn. In a computer simulation done by the planetary scientist Brett Gladman at the University of British Columbia, six million "rocks" were ejected from the Earth and sent on random, gravitationally determined trajectories around the Sun. Of those six million, only about a half dozen crashed into the surface of Europa. Slightly more make it onto the surface of Titan.

The rocks that do impact Europa do so at a speed that would cause them to vaporize on impact; none would be big enough to break a hole through Europa's ice shell. Therefore, any material that managed to survive the impact would be left on Europa's surface, exposed to harsh radiation. The energetic electrons and ions that pummel Europa's surface like rain from Jupiter's magnetic field would cook and kill any last surviving microbes.

In summary, it would be darn hard to seed Europa, or any of the ocean worlds of the outer solar system, with Earth life. Thus, even if we discovered DNA-based life there, we could reasonably conclude that those organisms represented a second, independent origin of life.

I should clarify that when it comes to looking for a separate, second origin of life and biochemistries that could be different from ours, I am not referring to what I call "weird life"—that is, life that does not use water as its primary solvent and carbon as its primary building block. We examine this topic in more detail when we explore Titan's surface, but for now, when I refer to "alternative biochemistries," I am still referring to water- and carbon-based life. What is "alternative" here is the prospect of finding different molecules that run the show, that is, an alternative to DNA.

In our efforts to see if biology works beyond Earth, we start with what we know works. Water- and carbon-based life *works* on Earth, and thus we look for similar environments beyond Earth.

But that is not to say that understanding the nature of water- and carbon-based life on Earth came easy. Earth's ocean has always been central to the story of life on Earth and how our planet balances ecosystems on a global scale. As Jacques-Yves Cousteau said at the beginning of his Ocean World book series, "The ocean is life." For millennia, the creatures of our ocean have populated our imaginations and guided our scientific pursuit of piecing together the tree of life on Earth.

OUR OWN ALIEN OCEAN

The story of the search for life beyond Earth is also the story of our growing understanding of our own oceans' depths, and our discovery of the secrets they hold. You may have seen old maps, maps where sea monsters, giant squid, and dragons dot the vast expanse of the seas yet to be explored. One globe from 1510 bears the phrase that has become synonymous with unknown dangers and risks: *Hic sunt dracones*; "Here be dragons."[2]

The ocean has long been the source of myths and legends. It was—and continues to be—home to aliens of a closer kind. How did we come to explore our own ocean and its many secrets?

The *Challenger* expedition, departing England December 1872 and returning home four years later, was the first to survey the biology of the deep ocean. The expedition's Royal Navy ship, the HMS *Challenger*, carried its crew around the world's oceans, covering a distance equivalent to a third of the way to our Moon. It was, and remains to this day, one of the most important and pioneering scientific expeditions to set sail.

Chief scientist of the expedition, Charles Wyville Thomson, was given permission from the Royal Navy to overhaul the ship, removing much of the weaponry on board and replacing it with instruments and labs. One instrument was little more than a fancy spool of line with a weight on the end. Simple as it was, this instrument would prove key to a great discovery.

In March 1875, the HMS *Challenger* was located southwest of Guam and dropped this line to a depth of 5.1 miles (8.2 km)—deeper than any line had been dropped into an ocean. Subsequent expeditions would reveal that the *Challenger* expedition had found the Earth's deepest ocean trench, the Mariana Trench, nearly 7 miles (11 km) at its deepest point.

The team on board the *Challenger* used nets and dredges to haul up whatever serendipity provided. Many of the creatures came up as gelatinous blobs, invertebrates whose form and function could be truly appreciated only in their native environment. The ethereal, alien beauty of a jellyfish, when hauled onto the deck of a ship, is reduced to colorless goo.

Long frustrated with their inability to directly observe creatures of the deep, explorers have worked throughout the ages to get firsthand access to the deep ocean. Diving bells were the original solution. If you have ever tipped over a canoe and swum underneath to breathe the air trapped by the canoe, you have experienced the basic operation of a diving bell. Imagine that a canoe has been weighted to sink to the bottom of a lake or river. The trapped pocket of air is the breathing space for anyone brave enough to take the trip down to the bottom.

According to paintings and reports, diving bell contraptions date as far back as Alexander the Great, a few centuries before the common era (BCE).[3] In a fun twist of the stars and seas, none other than Edmund Halley, discoverer of Halley's comet, innovated on the diving bell, creating a version in which the air could be cycled out and replaced with fresh air from the surface, carried by weighted canisters on a line.

In 1691, less than a decade after his observations of the comet that would come to bear his name, Halley and five colleagues descended in his diving bell to 60 feet (nearly 20 meters) in the River Thames. It was a small but significant step in getting humans deeper and opening our eyes to life in the depths.

The real leap in deep ocean exploration came in the late 1920s and early 1930s, when the engineering and science team of Otis Barton and William Beebe created and deployed their bathysphere—a hollow, steel sphere only 4 feet, 9 inches (1.5 meters) in diameter, with 3-inch-thick quartz windows. This sphere was connected to a cable on a ship's winch that could lower it down and haul it up. Electrical cables also enabled communication to the surface and provided power for lights.

In 1934 the pair of explorers—along with the support of their team from the New York Zoological Society, including naturalists Gloria Hollister and Jocelyn Crane, and engineer John Tee-Van—achieved their long-sought goal of reaching a depth of more than a half mile (nearly 1 km).

The team made many dives off the coast of Nonsuch Island in north Bermuda, and the creatures they saw filled catalogs of new and never-before-captured species. Beebe and the team were the first to study life in the deep ocean in its natural environment.

Beebe's description of a dive to 2,500 feet in early August 1934 captures his surreal experience: "There are certain nodes of emotion in a descent such as this, the first of which is the initial flash. This came at 670 feet, and it seemed to close a door upon the upper world. Green, the world-wide color of plants, had long since disappeared from our new cosmos, just as the last plants of the sea themselves had been left behind far overhead."[4]

On numerous occasions Beebe's writings and radio broadcasts linked the dark sea, peppered with bioluminescent creatures, to the twinkling stars of the night sky. After his successful dive with Barton to 3028 feet, Beebe wrote: "The only other place comparable to these marvelous nether regions, must surely be naked space itself, out far beyond atmosphere, between the stars, where sunlight has no grip upon the dust and rubbish of planetary air, where the blackness of space, the shining planets, comets, suns, and stars must really be closely akin to the world of life as it appears to the eyes of an awed human being, in the open ocean, one half mile down."[5]

The connection between sea and space appears time and again in exploration. Indeed, when NASA launched the first planetary spacecraft toward Venus in 1962, it was not given a name of astronomical significance but one that was connected to our ocean: *Mariner*.

And just two years before Mariner flew by Venus, humans themselves would make the plunge to the deepest part of the ocean for the first time, seven miles down in the Challenger Deep region of the Mariana Trench. In 1960 the *Trieste*, a 100-ton vehicle consisting of a sphere that fit two people (Jacques Piccard and Don Walsh) and a giant, buoyant carafe of gasoline, dropped to the deepest point in our ocean.

The dive of the *Trieste* bathyscaphe[6] marked what some hoped would be the beginning of an ambitious program to explore the deepest regions of our ocean. Designed by a Swiss inventor (Auguste Piccard, father of Jacques), built in its namesake region in Italy, and purchased by the United States Navy, it was the culmination of centuries of ocean exploration that sought to answer the question of what lies below not what lies above and beyond.

On that historic dive little was actually seen, as sediment that stirred up from the seafloor clouded much of the view, and Piccard and Walsh could not stay on the bottom for long. The deep ocean remained largely unseen.

But seventeen years after the *Trieste* landed in the Mariana Trench, in the spring of 1977, the abyss would give way to new insights into how life works in some of the most extreme environments on planet Earth. The veritable aliens within our own ocean would finally be revealed.

At that time, it was hard to imagine that there were still entire ecosystems on our planet yet to be discovered: the continents had been mapped; the poles had been reached; humans had touched down in the deepest point within Earth's ocean; the footprints of 12 humans even dotted the landscape of the Moon. What game-changing discoveries were left to be made?

Plenty, it turns out.

In that spring of 1977, a team of scientists set off to explore the Galápagos Rift, a region of the seafloor near the Galápagos Islands. They wanted to find out what was causing temperature anomalies in the region. Previous expeditions had measured these anomalies with instruments dropped down on cables and dragged around the ocean. The thinking at the time was that the plate tectonics of the spreading Galápagos Rift was creating a lot of localized heat; hot rocks were creating hot water, simple enough.

As part of the expedition, the team used *Alvin*, a US submersible, expecting to make important observations and discoveries about how geology works. But what they saw instead called into question how biology works.

At a depth of over 6,500 feet (2,000 meters), the lights on *Alvin* revealed structures that resembled tall and tortuous chimneys, billowing out "smoke" like active smelting plants from the Industrial Revolution. This was not smoke but clouds of fluids jetting out into the ocean at temperatures well beyond boiling—nearly 750 °F (400 °C). These fluids do not boil because they can't: the pressure is too high at those depths. These "superheated" fluids contain gases like hydrogen, methane, and hydrogen sulfide, as well as minerals that dissolve in the high-temperature and high-pressure fluids. The *Alvin* team had come across what we now call a hydrothermal vent—essentially a powerful, gushing hot spring at the bottom of the ocean.

The surprise was not so much the vents, but rather the bizarre and beautiful ecosystem surrounding the vent chimneys. Like a deep ocean version of animals congregating at a watering hole in the African savanna, the chimneys were host to never-before-seen creatures—red tube worms, stark white eel-like fish, and golden mounds of mussels—that were thriving in this extreme environment where conventional wisdom had said no animals should exist. And yet there they were.

How were these creatures surviving? What was sustaining this astonishing ecosystem?

On the surface of the Earth, the base of the food chain is driven by photosynthesis. Algae and plants harness the Sun's energy, breathing in carbon dioxide, extracting the carbon to build the structures of life, and then exhaling oxygen. Small organisms and animals eat the photosynthetic organisms, and then larger organisms eat those, and so on.

At the bottom of the ocean, however, the Sun is nowhere to be seen, and the food chain as we know it breaks down. Light from the Sun penetrates about 1,000 feet (300 meters) down, but beyond that, photosynthesis is not an option.

What was the base of the food chain at these hydrothermal vents? This is where the chemistry of the vents come in to play, offering essential nutrients and forming oases for life on the seafloor. The vents erupt hydrogen, methane, hydrogen sulfide, and a host of metals, many of which turn out to be tasty treats for microbes. The microbes utilize chemosynthesis instead of photosynthesis. Here the prefix "chemo" denotes that the microbes are synthesizing what they need for life with chemicals from the chimneys instead of photons from the Sun.

Chemosynthesis forms the base of the food chain at the vents. Microbes survive off the fluids and gases from the hydrothermal vents, and then larger organisms eat them, followed by larger creatures that eat those organisms, and so on. In some cases, the larger organisms were found to have also developed symbiotic relationships with the microbes—hosting the microbes within their bodies in exchange for the microbes detoxifying the water. All in all, a new and very surprising type of ecosystem had been discovered on that historic dive to the Galápagos Rift in 1977.

Life—large and small—was found to thrive in a region where most would have said it was not possible.

Only two years later, in March and July 1979, twin *Voyager* spacecraft would fly past Jupiter, capturing the first close-up images of Europa and Jupiter's other large moons. Those images would lay the foundation for thinking there might exist oceans of liquid water in a region where most would have said it was not possible.

In those brief years of the late 1970s, two seemingly disparate but phenomenal discoveries helped pave the way for a new approach to the search for life beyond Earth. The prospect of a liquid water ocean within Europa was all the more exciting once it became clear, through the discovery of the hydrothermal vents, that life could thrive in the dark regions of our ocean, cut off from the Sun, in a manner perhaps similar to that of an ice-covered ocean.

Our own alien ocean, hidden in the abyss, provided a glimmer of hope that distant oceans beyond Earth might also harbor life. In the chapters ahead, we dive deep into how we think we know these oceans beyond Earth exist and why we think they could be habitable. But first, we need to understand the sweet spot for habitability and why some of these ice-covered moons might reside in that sweet spot. To develop that understanding, we begin with the classic story of Goldilocks.

CHAPTER 2

THE NEW GOLDILOCKS

In the early days of planetary science and astronomy, our ideas about which worlds could possibly be habitable were largely biased by our own experience.

We live on a beautiful, blue world where a liquid water ocean covers much of the surface. Our ocean is in contact with a thick atmosphere full of oxygen and nitrogen. Plants grow in the rich soil and help sustain a complex global biosphere. We can sit on a beach, as the waves lap onto the shore, and look across the horizon to the large, glowing orb in the sky that makes it all possible—the Sun.

The Earth's specific proximity to the Sun is what makes our ocean possible: we're close enough to get the energy needed to sustain liquid water on the surface, but not so close that the water boils away. Earth resides in our solar system's sweet spot for sustaining liquid water on the surface of a planet.

Much of the early thinking about habitable worlds thus followed a star-centric logic—life requires liquid water; liquid water requires energy from a star; and therefore any habitable planet must be at just the right distance from its parent star, such that it receives enough energy to maintain a liquid water ocean on its surface. Get too close, or go too far, and the planet will be like Venus or Mars, respectively—any oceans they once had have long since boiled away or frozen. What a habitable world needs is to be like Earth, in the right place and at the right temperature, not too hot and not too cold—like that perfect bowl of porridge in the story of Goldilocks.

This Goldilocks-like distance from a planet's parent star has long been referred to as the "habitable zone" by astronomers and planetary scientists. It is an important concept that has helped guide our thinking about the thousands of exoplanets we now know exist around other stars. By figuring out the mass of these planets and how far they are from their parent star, we can begin to assess whether or not they could have liquid water on their surface.

For decades, we judged a planet's potential habitability according to this "Goldilocks" scenario of too hot, too cold, and just right, with Venus, Mars, and the Earth representing the little bowls of porridge that Goldilocks tastes, before the bears come home.

But recently, we've learned that there's more to the story. On the ice-covered moons of the outer solar system, we've discovered a new Goldilocks zone—a new way of determining if a world could be habitable.

It turns out that there's more than one way for a world to maintain a liquid water ocean. In this chapter, I will describe how the tidal tug of a moon as it orbits a planet can sustain liquid water oceans and how decay of radioactive heavy elements can contribute to the heat needed for maintaining liquid water.

And the Goldilocks requirements for liquid water is only one example of the conditions being "right" for life as we know it. Life needs three things: liquid water, the elements to build life, and the energy to sustain life. Each of these three parameters brings with it a Goldilocks condition of sorts. In this chapter we dive into the many ways in which planets and moons form and evolve into potentially habitable worlds.

THE SHAPE OF WATER

The physics makes sense but it's a bit unnerving nevertheless.

I'm riding on a snow mobile heading north across the Arctic Ocean. Our team is making its way to where the ice is starting to break apart as the season turns from spring to summer. The ice beneath us is probably four to eight feet thick, and beneath that the ocean is a few hundred feet deep. I can't stop imagining scenes of the ice breaking apart and all of us careening straight into the water and down to a chilly death.

Thankfully, the laws of physics don't care what my imagination has to say. We're perfectly safe here—in fact, if the ocean were perfectly calm, it would take only a foot or so of ice to support the weight of me and my snow mobile. Just a few feet of ice would make for a fine ride atop the Arctic Ocean. Although we tend to think of it as precarious and breakable, ice turns out to be a remarkably robust barrier—not just for me, but also for what lies beneath. The ice covering the ocean keeps us from falling in, but it also serves as a blanket for the ocean below, keeping in heat that would otherwise be lost to the cold air above.

Before we dive into the more complicated topics associated with the new Goldilocks scenarios for habitability, it is important to appreciate the simple beauty of the water molecule and the changes that occur as it transforms from liquid to solid. Without these slight changes, it's game over for any ocean world. All water would simply freeze to ice.

Water is an amazingly elegant substance. When exposed to a cold environment, it naturally forms a protective insulating barrier that, floating atop it, shields the liquid water from further exposure to the cold. It's such a commonplace sight that it's easy to take for granted, but this simple fact of physics may be the key to the largest volume of habitable real estate in our universe.

Consider a thought experiment: imagine that ice sank. If it did, then water would behave much like butter. When you melt a few sticks of butter in a bowl and let them cool again, you see that solid pieces form on the surface and then sink to the bottom. The liquid butter continues to cool on the surface and sink to the bottom until all of the liquid butter has solidified. Now imagine a lake of liquid butter in the winter. It would simply freeze solid because the solid butter sinks and the remaining liquid butter is left exposed to the cold.

Fortunately, lakes aren't made of butter, and ice doesn't sink—it floats. The surface of a lake will freeze in the winter, and the ice that forms gets thicker throughout the season. As it gets thicker, it forms an insulating layer that prevents the cold from seeping deeper, making it harder and harder for the depths of the lake to freeze. This is why lakes in the Arctic that are just a couple of meters deep can retain some liquid beneath the ice, providing a safe home to fish and all sorts of other creatures. This is also why billions of habitable liquid water oceans may exist throughout our universe.

But *why* does ice float? Again, this is something we typically take for granted. Ice floats due to a very simple, but important, change in the geometry of bonding between the water molecules. That change in geometry, as water goes from liquid to solid, generates a larger-volume, lower-density material and increases the entropy of the system.

The density change in water across the liquid–solid boundary (at the pressure of Earth's surface) goes from 0.9999 grams per cubic centimeter (g/cm^3) of liquid water at 0 °C, to 0.9167 g/cm^3 once it freezes at 0 °C. That is about a 9% change in density! Curiously, water has an even higher density at 4 °C than at 0 °C in the liquid phase. At 4 °C water is 1.0000 g/cm^3.

This contrasts with the behavior of most rocks. For example, when lava from a volcano cools, it becomes denser. As it solidifies, basalt lavas gradually increase in density from about 2.6 g/cm^3 at the melting temperature of roughly 1,200 °C, to 2.7 g/cm^3 in solid rock form at 1,000 °C. Instead of floating, the solid rock sinks.

The density change in water results from a slight change in the angle between molecules linked together in the crystal structure of ice. Water (H_2O) is made up of one oxygen atom and two hydrogen atoms. The oxygen atom has a total of eight electrons: two buzz around in the inner electron shell and six are in the outer electron shell. There is no perfect analogy for how to think about these electron shells, but for the sake of understanding the water molecule, just imagine that these shells are like nested Ferris wheels. Each seat on the Ferris wheel can house two electrons. The inner shell—or inner Ferris wheel—has only one seat and the outer shell has four seats. The outer shell could host a total of eight electrons (four seats, two electrons per seat), but the oxygen atom has only six electrons left after using two to fill the inner shell. The oxygen atom wants two more electrons so it can fill those two empty seats in its outer Ferris wheel shell.

Hydrogen atoms, meanwhile, are simple Ferris wheels—one seat going around and around, and one electron in that seat for two. Hydrogen atoms are happy to share their single electron with any atom that will also share an electron with them. For the water molecule, the net result is that two hydrogen atoms team up with one oxygen atom; each hydrogen shares its one electron, while oxygen shares its outer shell electrons. The Ferris wheels of oxygen and hydrogen become interlocked, and they share passengers so that all the seats in all the Ferris wheels appear full. Everybody makes out well in this shared electron arrangement. This is the basic principle behind covalent bonding in chemistry (Figure 2.1a).

But the plot thickens. The hydrogen atoms in the water molecule become slightly positively charged because their electrons are on a partial loan to the oxygen atom, leaving the positive charge of the proton in each hydrogen atom to be slightly stronger.

Meanwhile, oxygen now has ten electrons around it, but still only eight protons in its nucleus. The oxygen atom has an excess of negative charge because it is borrowing two additional electrons to fill its outer shell.

As a result of this sharing arrangement, the water molecule becomes "polar," meaning that it has a region with a slight positive charge (the hydrogen ends) and a region with a slight negative charge (the oxygen region). The negatively charged oxygen end of one water molecule attracts the positively charged hydrogen end of another. Molecules of water are constantly sticking to each other and breaking apart, like miniscule magnets being shaken around. In chemistry, this kind of bond is called a hydrogen bond and is largely responsible for many of the interesting properties of water, including the density differences.

Digging a little deeper, we can begin to see how the hydrogen bond results in ice floating. Going back to the Ferris wheel analogy, each seat on the Ferris wheels of the water molecule like to spread themselves out so they are far away from the other seats. This leads to water molecules forming a bit of a pyramid, or tetrahedron (Figure 2.1b).

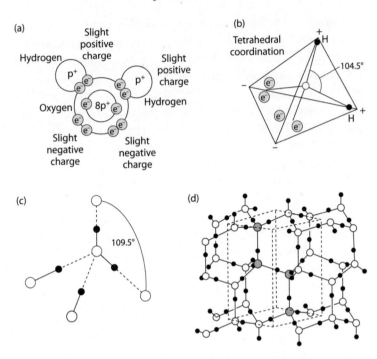

FIGURE 2.1. Atomic structure of a water molecule and covalent bonding between water molecules. (a) The oxygen atom in water has eight electrons—two in the inner electron shell and six in the outer shell. Hydrogen atoms have only one electron in a shell than can accommodate two electrons. The result is that two hydrogen atoms team up with one oxygen atom to form a water molecule. (b) As a result of electron sharing, the two hydrogen atoms have a slight positive charge, and the oxygen end of the molecule has a slight negative charge. The arrangement of these charges leads to a tetrahedron, or pyramid-like shape, for charge around the molecule. (c) Due to the positive and negative regions on each water molecule, they link together with the hydrogens of one molecule pointed toward the oxygen atoms of neighboring molecules. Similarly, the oxygen in one molecule attracts two hydrogen ends from neighboring molecules. (d) The charge and arrangement of water molecules leads to a hexagonal structure, which is ultimately why snowflakes always have six points.

The angle between the hydrogen branches of the pyramid (i.e., linking H-O-H) is 104.5 degrees (Figure 2.1c). The electrons distributed on the other side of the molecule also form an angle of 104.5 degrees, with the oxygen atom at the vertex. For a water molecule on its own, this is the end of the story. It forms a nice little compact, slightly distorted pyramid.

If we now take lots of water molecules and mix them together, they will link together, via the hydrogen bond. In liquid water, molecules can be on their own, or in chains of linked molecules, or in sheets of linked molecules. (The surface tension of liquid water is a result of molecules linking together through hydrogen bonds.)

As the temperature drops and the molecules bounce around less and less, they get locked into more and more hydrogen bonds; the crystal lattice of ice begins to form (Figure 2.1d). Critically, when six water molecules link together, they form a hexagon. It's a slightly bent hexagon, with the angle connecting neighboring oxygen atoms spanning 109.5 degrees (i.e., linking O-O-O).

The hexagonal crystal lattice of ice, structured with this repeating angle of 109.5 degrees, takes up slightly more volume than the individual molecules with their span of 104.5 degrees. Freezing into ice causes the individual molecules to stretch slightly as they lock in with the surrounding molecules. The distance between molecules, which is a result of the hydrogen bonds, is slightly larger than the average distance when the molecules are in the liquid phase and are able to pack slightly closer together. It's a small but significant effect. This is what causes ice to float.

As an additional note of appreciation for the 109.5 degree angle and hexagonal crystal lattice of ice, consider those magical crystals that fall from the sky: snowflakes. All snowflakes have six sides—no more, no less—and it's due to the effect described above. All the amazing patterns we see in snowflakes are derived from the 109.5 degree angle of the hydrogen bond in water (Figure 2.2).

That ice floats is just one of water's wonderful properties. Another important attribute is that ice is a good insulator. Due in part to the hydrogen bond, ice and snow have low thermal conductivities, thus enabling ice-covered moons to retain the heat of their interiors and maintain water in a liquid state. If ice had a very high conductivity, any heat generated in the interior of worlds like Europa and Enceladus would pass rapidly out through the ice and be lost to space, leaving them frozen solid.

The simple facts that ice floats and conducts heat poorly are critical to why ocean worlds can, and do, exist.

FIGURE 2.2. The art of snow. Images of snowflakes captured by the pioneering photographer Wilson Bentley. Their beautiful hexagonal symmetry is a result the hexagonal structure of water molecules.

A GOLDILOCKS ZONE FOR TIDES

The properties of ice help explain how ocean worlds retain heat, but we haven't yet answered the question of where the heat actually comes from.

The source of the heat that makes these oceans possible is a game changer for habitable worlds. It is the truly new Goldilocks condition that moves us away from the constraints of the traditional habitable zone, defined by a star's energy and a planet's distance from the star. The new Goldilocks requirements move us into a realm of wide-ranging possibilities for creating and sustaining liquid water oceans. The source of energy in this new Goldilocks scenario comes from tides.

If an ice-covered moon orbiting a giant planet contains an alien ocean, then the heat for that world is likely generated by tidal energy. Tidal energy is a consequence of a world's gravitational interaction with another massive object, such as a planet or moon. As the two bodies move relative to one another, the world's solid mass actually stretches and relaxes because of the tug of its tides, like a rubber ball being squeezed again and again. If you squeeze and release a rubber ball dozens of times, it will start to heat up from all of the internal friction. Similarly, the tug of tides creates mechanical energy and friction within the object—which, in turn, creates heat. Hereafter we'll only consider the case of tidal heating in *moons* since that is most relevant to ocean worlds.

Two of the most important considerations for tidal heating are (1) the difference in the gravitational force across a moon, and (2) the change in gravity as a moon moves around a planet in its elliptical orbit. Two other critical parameters are the mass of the planet around which the moon orbits, and the period, or time it takes a moon to complete its orbit (which, as Johannes Kepler taught us, is a function of its distance from the planet). In the analogy to squeezing a rubber ball, these parameters can all basically be summarized as how intense is the squeezing, and how often does the squeeze and relaxation process occur?

Mathematically, the gravitational force between two bodies is proportional to the product of the masses of the two bodies divided by the square of the distance between them: $F = GMm/r^2$. Here G is the gravitational constant, M is the mass of Jupiter (or the Earth if we are considering our Moon), m is the mass of the moon being considered, and r is the distance between the two objects. We're not going to worry about the additional mathematical details of the gravitational interaction, but you may have seen the equation above in a physics class and it's good to appreciate that the physics of tidal interactions is simply an extension of this basic equation. What is critical is the interplay between a moon's eccentricity (which is a measure of the elliptical shape of its orbit), the mass of the planet it orbits, the distance of the moon from its planet, and the time over which it makes an orbit. The distance across a moon (its diameter) determines the differential gravitational stress across the moon, and the change in distance due to eccentricity determines how the strength of the planet's gravity changes as the moon moves through its orbit. These factors all come together to influence the net tidal heating within a moon.

At this point, it may be useful to put our own Earth–Moon system in context. We see the tides rising and falling on the shores of our ocean. Are tides a significant source of heating for the Earth? In short, no. The Earth and Moon orbit each other in nearly circular orbits, and they are relatively small, low-gravity objects, at least by planetary standards. What this means is that, as they orbit each other, the distance between them doesn't change much—thus, the gravitational field doesn't change, and the tides don't either. Fixed and constant, they generate almost no heat from their motion. Think of that rubber ball, squished in your hand, but you never let go. It's deformed, but no heat is being generated. Without repeated squeezing and releasing, the ball doesn't heat up from changing shape. In order for a moon or planet to experience significant tidal heating it must experience a *changing* gravitational field, which is like your hand squeezing and letting go of that rubber ball.

The little tidal heating that does occur as the Earth and Moon orbit each other results from the solid part of Earth's surface—continents and seafloors—rising and falling ever so slightly. The rocky part of the Earth rises and falls by a few centimeters to as many as 25 centimeters, depending on the alignment of the Earth, Moon, and Sun. The amount of heat produced from tides is just a few milliwatts per square meter (mW/m^2), which is negligible compared to the 1,388 W/m^2 we get from the Sun.

Tides do not create much heat for planet Earth or the Moon, but they obviously play a very significant role in the dynamics of our ocean. The motion of our ocean's liquid water creates very little heat. Unlike rocks that resist deformation but eventually bend and stretch, and in so doing create friction and heat, liquid water just flows to accommodate the gravity of the tides. There is no significant heating because water moves freely and does not resist the motion of the tides.

The Moon's gravity raises a tidal bulge of ocean water on the Earth directly beneath the Moon's position, and there is a corresponding bulge of water on the opposite side of the Earth. It's pretty intuitive that there should be a high tide bulge under the Moon, but why is there one on the other side of the Earth?

Recall that tides are caused by the difference in the gravitational force across a body. Since the force of gravity is inversely proportional to the square of the distance between two objects, the Moon tugs the Earth about 6% more strongly on the side that faces it than it does on the side farthest away. The Earth itself gets pulled closer to the Moon, leaving behind the water on the far side of the Earth, thus creating a high tide there.

The Earth rotates more quickly (one full rotation in 24 hours) than the Moon revolves around it (27.3 days), and thus the Moon "sees" a different region of the Earth all the time. The high tide bulges of ocean water stay aligned with the Moon, and the solid Earth rotates through them (Figure 2.3).

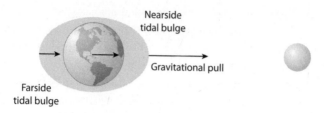

FIGURE 2.3. The tug of tides. While the rocky Earth rotates, the tides "stay" in the same position. The gravitational tug of the Moon pulls an ocean bulge toward the Moon on the near side of the Earth. The tidal bulge on the opposite side of the Earth is a result of the water that is left behind because the gravitational pull from the Moon is lower on the far side of the Earth.

Since there are two bulges, and the Earth completes its rotation in 24 hours, every place on Earth passes through two high tides and two low tides every day. This fixedness of the tides with respect to the Moon is critical. The tides on Earth are not really rising and falling, and so they do not, in fact, create a lot of mechanical energy.

While our Earth–Moon tidal dynamics do not create much internal heating, numerous moons of the outer solar system undergo very significant heating from tides. The key differences are as follows. First, these moons orbit giant planets that have very strong gravitational fields, within which the moons orbit. Second, if a moon has an elliptical orbit around a planet, then the distance between it and the planet is always changing and, therefore, so is the gravitational field. The tidal bulges will increase and decrease in size as the moon goes from its closest approach (periapse) to its most remote position (apoapse).

With each orbit, the rise and fall of these tidal bulges causes stretching and relaxation, which creates friction and ultimately heat. Many of the ocean worlds of the outer solar system, including Europa, Ganymede, an Enceladus, have orbits that are elliptical and that cause this kind of tidal stretching and heating.

Jupiter's four large moons provide a useful case study in this new Goldilocks framework for tidal heating. By studying the orbits of these moons, the potential for this new Goldilocks zone emerged. In 1979, just prior to the arrival of the *Voyager* spacecraft at Jupiter, Stanton Peale and

Patrick Cassen (from the University of California Santa Barbara) and Ray Reynolds (from NASA Ames Research Center) published an article on the theory of tidal energy dissipation in the Jovian system.[1] Remarkably, they concluded that tides could cause much of the interior of Io, Jupiter's innermost moon, to be molten. Io could therefore be volcanically active. They even made the bold prediction that images from the *Voyager 1* spacecraft might reveal such activity.

Peale, Cassen, and Reynolds made one of the most elegant and exciting predictions followed by discovery in the history of planetary science. Soon after their article was published, *Voyager 1* flew by Jupiter and returned stunning images of volcanic plumes erupting from Io into space.

Not long after, the team published a similar article about the effect of tides within Europa. The title of their article, "Is there liquid water on Europa?," was provocative.[2] This was the first time that a scientifically rigorous and robust mathematical argument had been made for an ocean existing within Europa. It was, in my opinion, the birth of the "New Goldilocks" conditions—a new way to determine a possible habitable zone.

With the *Voyager 1* and 2 flybys, and the subsequent exploration of the Jovian moons with the *Galileo* spacecraft, we would come to understand the true power of tidal energy dissipation and this new Goldilocks requirement. For example, Io does not simply have volcanoes, it is *the* most volcanically active body in the solar system, even more volcanically active than the Earth. Volcanoes are erupting on Io *right now*.

Because Io has no atmosphere, the eruptions spew plumes of gas and lava out into space, forming umbrella-like shapes. Io's spectacular volcanic activity is the result of the tidal stretching and deformation it experiences as it moves along its eccentric (that is, not concentric) orbit around Jupiter, which is 318 times as massive as the Earth.

Io is made of rock and has an iron-rich core, and its rocky mantle is perfectly conducive to tidal heating. Tidal heating generates 2,400 W/m^2 on Io's surface—over 1,000 watts more than the incoming solar energy flux received by the Earth, and nearly equal to the 2,600 W/m^2 that Venus gets from the Sun! This is much too hot for a lot of water to exist on the surface. In this new Goldilocks scenario, Io is analogous to Venus: it has too much tidal energy and has lost almost all of its water. As a result, Io has lots of heat but no ocean for life.

Jupiter has dozens of moons, and of the four largest ones, Callisto is the farthest out. Callisto does have an ocean, but it's trapped beneath a very thick, old ice shell, likely sustained through the decay of radiogenic heavy elements in the moon's interior. Callisto experiences very little tidal heating. Although its orbit is highly elliptical—more so even than Io's—it is simply too far from Jupiter for this eccentricity to generate significant stretching and straining. Callisto's higher eccentricity (0.0074 to Io's 0.0041; where a value of 0 is a circular orbit) is largely offset by its greater distance from Jupiter.

In the new Goldilocks framework, Callisto is akin to Mars. It was, and perhaps still is, habitable, but the tidal energy dissipation is very small; as a result, Callisto is colder and less active than the large inner moons.

In between Io and Callisto lie Ganymede and Europa—occupying the sweet spot of the new Goldilocks zone. These two moons are stretched and squeezed by tides enough to generate tens to hundreds of milliwatts per square meter of internal heating. In the case of Europa, this is enough to maintain a liquid water ocean of approximately 100 km (60 miles) in depth, with a rocky seafloor—perhaps dotted with hydrothermal vents—all of which is overlaid by a relatively thin ice shell (a few kilometers thick to as much as 30 km thick). This ice shell is thin enough that its surface chemistry may provide a window into the chemistry—and possibly the biology—of the ocean below.

Europa and Ganymede are also the beneficiaries of a curious property of the Jovian system that keeps their orbits elliptical and helps maintain their tidal heating. The three innermost large moons—Io, Europa, and Ganymede—are like three kids on a swing set whose back-and-forth motion has gradually synchronized into a pattern. For every one orbit that Ganymede makes around Jupiter, Europa makes two; and for every one orbit that Europa makes, Io makes two. Ganymede, Europa, and Io's orbital periods are thus locked in a 1:2:4 ratio, known as the Laplace resonance (after the French mathematician Pierre-Simon Laplace, who discovered it in the early 1800s).

The Laplace resonance is important because it forces each moon to stay in an elliptical orbit. Typically, over time, orbits circularize and lose their eccentricity (i.e., they become less elliptical). But in the Jovian system, the three inner large moons regularly align in pairs: Io–Europa, Europa–Ganymede, and Io–Ganymede. When this happens, the aligned moons tug on each other, leading to a "forced eccentricity" and causing their orbits to each stay slightly stretched into an ellipse instead of becoming perfectly circular. (Note that all three never line up together on the same side of Jupiter.)

Although the exact timing of the start of the Laplace resonance around Jupiter remains a topic of considerable study, at some point in the distant past (perhaps billions of years ago), Io began to retreat from Jupiter. Gradually, it got close enough to Europa to exert some gravitational influence on it. These two moons then engaged in a two-body resonance, systematically tugging on each other to create forced eccentricities and possibly settling into the 1:2 resonance they experience today.

Over time, both moons continued to lose energy and momentum to Jupiter, and their orbits grew larger. Eventually, they got close enough to Ganymede to begin to influence its orbit. The tug between the three moons then stabilized into the 1:2:4 resonance that we observe.

One day, Callisto will be part of this clockwork too. As the three innermost large moons continue to retreat outward, they will—perhaps hundreds of millions of years from now—expand far enough out to influence Callisto's orbit. Will there then be a complete resonance of 1:2:4:8? It's hard to predict since so many different factors influence energy loss and momentum transfer, such as tidal energy dissipation or how Jupiter itself responds to the interaction.

NASA's *Juno* spacecraft, which began orbiting Jupiter in 2016, is probing some of these big picture questions about planetary dynamics and interiors. Once we have a clearer picture of how Jupiter works, we'll be able to better understand its relationship to the ocean worlds trapped in its orbit.

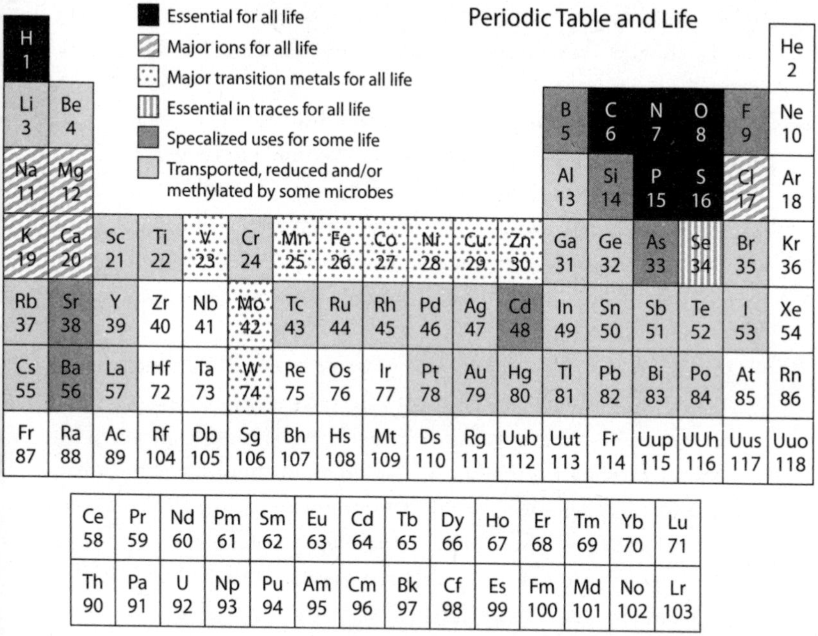

FIGURE 2.4. The periodic table and a grayscale code indicating which elements are essential for life, which play a major role in life, and which are used in specialized cases for some life. (Adapted from Wackett, L. P., Dodge, A. G., & Ellis, L. B. (2004). Microbial genomics and the periodic table. *Appl. Environ. Microbiol.*, 70(2), 647–655)

GOLDILOCKS GETS GEOCHEMICAL

Life as we know it requires several basic elements, which form the compounds on which life depends. These elements are carbon, hydrogen, nitrogen, oxygen, phosphorus, and sulfur—collectively known by the acronym CHNOPS. In addition to these elements, life uses a smattering of some 48 other elements (Figure 2.4).

These building blocks are, naturally, critical to the construction of any habitable world. However, when you look where they occur in great abundance in our solar system, it turns out that the best real estate isn't actually our own familiar planet, but rather it is in the outer reaches of the solar system. Years ago, my PhD advisor, Chris Chyba, liked to articulate this point by saying, "Earth is a bad place for life." That is, for all the spectacular abundance of life on our planet, the raw materials to create it are actually hard to come by.

When it formed, the Earth did a great job amassing the heavier elements that life needs: iron, magnesium, sodium, potassium, calcium, nickel, copper. All of these can be found in abundance in the rocks on Earth. But when it comes to the lighter elements—specifically the CHNOPS elements—Earth didn't do so well. The Earth has about 1% of the carbon and nitrogen available in the outer solar system, where Jupiter and Saturn roam.[3]

Our home has only a thin veneer of water and carbon, those two ingredients most essential to life. What's more, a significant fraction of the water it does have may actually be an accident, as it were, delivered by icy comets after Earth's formation. The situation elsewhere in the inner solar system is even worse: bone-dry and rocky. Although Venus and Mars both likely had oceans once, the amount of water in them (perhaps also delivered via comets) would have been a proverbial drop in the bucket, even in their heyday. The problem is that these planets were too close to the Sun when they formed. When the solar system was first coalescing, the inner solar system was largely "baked-out." It was too hot for light compounds (often called volatiles) to stick around—literally. Because of the high heat, it was impossible for molecules of water, methane, ammonia, and sulfide to "stick" together, or to anything else, and condense into liquids and solids. Remaining in the gas phase, they were essentially blown away—to the outer reaches of our solar system, where they could condense into ices and other materials (Figure 2.5).

In planetary science, we call the particular regions where volatiles tend to condense from gas to solid the "snow lines" or "frost lines" for various materials. The snow line for water occurs at about the distance of the asteroid belt, which is roughly three astronomical units (AU) from the Sun. (One astronomical unit is defined as the distance between the Earth and the Sun.) The snow lines for methane, carbon dioxide, and sulfide (H_2S) are all beyond 5 AU, which is about the distance to Jupiter. Early on, when the Sun was still settling down and reaching a stable temperature, these lines were likely at slightly different distances. But the

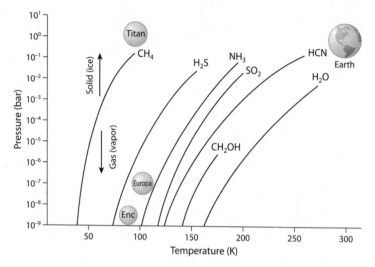

FIGURE 2.5. Sublimation curves for a variety of molecules. Above the curves, the pressure is high enough for solid ice to form; below the curves the pressure is too low and the molecules stay in gas phase. The temperatures and pressures of Titan, Europa, and Enceladus freeze molecules into ices and make these worlds rich with life's key elements. (Adapted from Fray, N., & Schmitt, B. (2009). Sublimation of ices of astrophysical interest: A bibliographic review. *Planetary and Space Science, 57*(14–15), 2053–2080)

important point is that they have always been far beyond the inner solar system.

Thus, when the planets and moons formed, they formed from the materials in their region. The story is complicated—especially if you consider the possibility that giant planets like Jupiter might have migrated in close to the Sun and then back out to its current distance—but for the most part, the inner solar system formed from rocky materials, and the outer planets formed from the lighter materials (e.g., gas and ices) that had drifted farther out. In general, the CHNOPS elements formed volatile compounds that condensed into ices in the outer solar system and that formed the foundations for the planets and moons of that region.

The densities of the planets in our solar system reflect, in part, the relative distributions of these materials when they formed. Although the inner planets are smaller, they are much denser than the gas and ice giants of

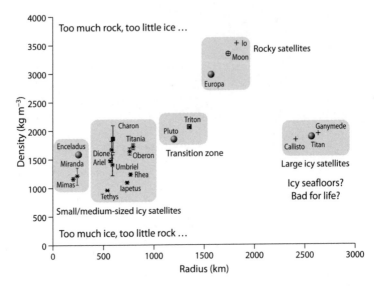

FIGURE 2.6. A balance of water and rock may be critical to habitability and the origin of life itself. For rocky and dense moons, like our Moon and Jupiter's Io, there is no liquid water, which is bad from a habitability standpoint. Larger moons like Callisto, Titan, and Ganymede may run into problems of heavier phases of ice forming their seafloors, possibly isolating liquid water oceans from any rocks in the interior. Worlds like Europa and Enceladus may occupy a sweet spot of size and density, such that liquid water can flow through rocky, chemically rich seafloors capable of powering life in their dark oceans. (Adapted from Hussmann, H., Sohl, F., Spohn, T. (2006). Subsurface oceans and deep interiors of medium-sized outer planet satellites and large trans-neptunian objects. *Icarus*, 185(1), 258–273.)

the outer solar system by virtue of the rocky materials that make them up. The gas and ices that compose Jupiter, Saturn, Uranus, and Neptune, meanwhile, make them very large, but low-density planets. The density of a body can serve as a useful indicator of its composition.

By looking at the densities of the moons of the outer solar system, then, we can get an estimate of the availability of CHNOPS elements and the other 48 elements of life (Figure 2.6). Most of the small and medium-sized moons of the outer solar system have low densities (less than 1,800 kg/m³), and thus primarily consist of icy materials. Meanwhile, Io, Europa, and our Moon have significantly higher densities and thus a larger fraction of rock than many of the others. Lastly, there's a cluster of moons (Titan, Ganymede, and Callisto) that are low density but also quite large.

This plot may be a map of another type of Goldilocks zone for habitable ocean worlds—one based on composition. On the one hand, as we've discussed, a habitable ocean world would need plenty of low-density ice and frozen volatiles to furnish the liquid water and carbon necessary to build life. On the other hand, it also needs rocky material for that liquid water to interact with, leaching essential heavier elements, like sodium, magnesium, iron, and nickel. Water–rock interactions are critical for creating the chemically rich soup that life needs, and accordingly, a habitable ocean world needs just the right balance.

Evidence suggests, for example, that Saturn's moon Mimas may have an ocean. But its density, at 1,150 kg/m^3, is so close to that of water (1,000 kg/m^3) that it probably doesn't have much rock, which has densities in the range of 3,000–8,000 kg/m^3, depending on the amount of iron and other heavy elements in the rocks. Mimas likely lacks enough water–rock interaction to build and power life. Conversely, Enceladus, at about 1600 kg/m^3, suggests both a lot of water and a significant rock fraction that can supply the other elements needed by life. As we'll see later, there is also good evidence from Enceladus's plumes that there is vigorous water–rock interaction happening on its seafloor.

Europa, meanwhile, appearing in the higher-density region of Figure 2.6, is perhaps curiously dense to be considered a leading candidate. Its density is about 3,000 kg/m^3, which is not too different from our dry and rocky Moon and from Io. Does this mean that Europa has too much rock and not enough water and light compounds? In short, no. Europa's water (including ice and liquid water combined) makes up 6% of its total mass, so Europa definitely has enough water. (For comparison, Earth's ocean amounts to only 0.02% of Earth's total mass.) Of the other CHNOPS elements, Europa also has a lot of sulfur, which we can see in the yellowish-reddish-brown regions on the surface. Some of that sulfur may precipitate out of Europa's ocean, but some of it arrives via those massive volcanic eruptions on Io.

Carbon and nitrogen, however, are a bit more mysterious. Results from the *Galileo* spacecraft show carbon dioxide ice at about 400 parts per million, which is comparable to the amount of carbon dioxide in our atmosphere. Thus we know that carbon exists and is at least available on Europa. The carbon dioxide that we see as ice on Europa's surface may be from gases dissolved in the ocean below, or it could be made on the surface from the radiation bombardment of organics or other carbon compounds. At this point, we just don't know.

The story for nitrogen is a bit more complex. We do not yet have any direct evidence indicating that nitrogen exists on, or within, Europa. That said, it must be there, given the abundance of nitrogen in the cloud of material from which Europa was formed. Nitrogen has been observed on Jupiter's outermost large moon, Callisto, and that gives us at least some reason to predict that nitrogen is on Europa too. Because the *Galileo* spacecraft provided more detailed data about Callisto's composition, it's not really a surprise that nitrogen was observed on Callisto but not Europa.

Interestingly, when we consider the snow lines for carbon and nitrogen species, like methane and ammonia, we see that the moons of Jupiter likely formed too close to the Sun for those species to condense out during formation. This may be, in part, why we do not see a lot more of these species on the Galilean moons. Out at Saturn's distance, however, it was perhaps cold enough to condense out these compounds. Both Titan and Enceladus, which we examine later, have a considerable amount of nitrogen and carbon.

Finally, there is one more plot twist to the geochemical Goldilocks. The size of the world also matters. If a moon is too large, then there is a chance that the pressure within that moon may be high enough for ice to form at the bottom of the ocean, potentially covering up a rocky seafloor and limiting the geochemistry—meaning hydrothermal vents and other geochemical cauldrons for life might not exist. Under very high pressure—about *twice* that found in the deepest part of Earth's ocean—cold water will form into a slightly different crystalline structure that is more densely packed than the ice that forms ice shells and floats. This denser ice sinks in liquid water.

Titan, Ganymede, and Callisto fall into this category. These moons are dense enough that they must have some rocky component, but they are also very large and their seafloors could be ice instead of rock. Ganymede, the largest moon in our solar system, is larger than the planet Mercury and even has a molten iron core that generates a magnetic field. Therefore, the size of these moons may create an odd bottleneck for life because they generate high pressures (in their interiors and at the bottom of their oceans) capable of forming this dense phase of ice.

Thus, it may be that the smaller, medium- to high-density moons like Enceladus and Europa have just the right density *and* size for chemically rich liquid water oceans in contact with rocky seafloors.

———

The new Goldilocks model is primarily about breaking away from the old model of habitability where the proximity of a world to its parent star was the primary factor in determining whether or not it could have a liquid water ocean and, ultimately, life. The heat generated from tidal energy dissipation breaks us away from that old model. Oceans can exist far away from the heat of a star; the dance of a moon around a giant planet is more than enough to sustain vast, global liquid water oceans.

Along with the new Goldilocks constraints for tides, we have, in this chapter, covered several other factors that make the ice-covered moons of the outer solar system particularly well-suited for hosting habitable oceans. The outer solar system was cold enough to condense out key elements for life as we know it, such as carbon, nitrogen, and sulfur, but rich enough in the heavy elements to ensure that at least some of these moons might be able to sustain the chemistry needed to power and build life. This combination, I argue, creates the ultimate Goldilocks scenario for habitable ocean worlds beyond Earth.

Three worlds emerge as the top candidates for occupying this new Goldilocks sweet spot. Europa and Enceladus may have the right combination of liquid water, elements, and energy needed to give rise to life and power life as we know it. Lastly, Titan, although perhaps too large to have a rocky seafloor, is flush with carbon and interesting organic chemistry that make it hard to resist when it comes to the prospect of life.

In the chapters that follow, we first gain a better understanding of how we think we know these oceans exist, using Europa as an example, and then we go into detail about Enceladus and Titan. We then briefly explore many of the other possible ocean worlds in our solar system before turning to the topic of habitability and life's origins in more detail.

PART II

DISCOVERING AN OCEAN IN
THREE EASY PIECES

CHAPTER 3

THE RAINBOW CONNECTION

In the first few chapters we focused on the "why" of ocean worlds. Why do they exist and why are they able to maintain potentially habitable oceans? Tides and the variety of Goldilocks sweet spots help to answer the "why."

But we have yet to address the "how." How do we know these oceans might exist?

In this chapter, and the two that follow, we dissect the science behind how we think we know these oceans exist. There are many different clues from many different lines of evidence. It's a bit of a detective story, and it involves telescopes on the ground and spacecraft careening by the worlds at perilously close distances. Europa, Enceladus, Titan, and many of the other potential ocean worlds share similar themes in the discovery process. I will use Europa as the example and describe the science in great detail of how we think we know Europa has an ocean, but the techniques apply to our understanding of many of the ocean worlds. Europa is perhaps the best starting point because it was the first moon for which we gathered evidence of a subsurface ocean, and it is also the one for which we have the most mature scientific understanding.

I break down the science of "how" we know into three easy pieces. The first piece involves making a rainbow connection. The other two pieces involve babysitting a spacecraft and learning to love airport security and are described in the next two chapters.

————

Rainbows are physics and chemistry, art and beauty, at their best. Rainbows are also the first piece of the puzzle that helped unveil Europa's subsurface ocean.

When you see a rainbow in the sky, what you are seeing is light from the Sun passing through our atmosphere and through tiny water droplets that spread out the different wavelengths of light into the colors we can see. Each water droplet acts like a little prism.

The many different photons of light from the Sun take slightly different paths through the raindrops. Photons of red light take the shortest path, while photons of violet light are bent onto a longer path. The resulting fan of photons separated by wavelength (i.e., color) creates that beautiful ribbon of color in the sky.

Contained within a rainbow is information about the composition of the original source of light (the Sun), the composition of Earth's atmosphere, and the composition of the raindrops through which the light passed. If you take a rainbow and tilt it on its side, making a chart of color versus intensity (the strength of that color relative to the other colors), you get a spectrum. The variations of intensity can help reveal which elements and molecules are contained within everything through which the light passed.

Spectroscopy is really quite profound. It allows us to figure out what something is made of without actually touching the thing itself. If an object is either emitting light directly or reflecting light (whether from the Sun or from a light bulb in the lab), we can measure the spectrum of that object and determine many of the materials from which it is made. Knowledge is acquired from a distance. For the most part, nothing is destroyed in the process of making the measurement.

Acquiring knowledge at a distance by using spectroscopy has, for good reason, been essential to the field of astronomy. We cannot fly to other stars to sample them directly, so we rely on spectroscopy to determine their compositions. Although our robotic spacecraft are amazing and have explored a variety of places in the solar system, our telescopes on Earth, equipped with spectrometers, are really what paved the way for figuring out the composition of the planets and moons.

Spectroscopy, as we explore in this chapter, enabled humans to determine that Europa's surface is covered with water ice—the first piece of the puzzle toward discovering an ocean below. But how did we mildly talented apes come to develop and appreciate the power of spectroscopy? How did we develop this tool that has enabled us to measure the chemical composition of anything from a rock sample in the lab to the surface of a distant moon?

THE LUX VERITATIS AND SIX LAMPS

Most people know that Galileo was the first to turn the telescope toward the night sky, but few know who was the first to examine the light spectrum of an object in space. Although lesser known, this development was profoundly important—it revealed that chemistry, and the principles of chemistry, work not only on Earth but also on worlds and wonders beyond Earth. The distant marvels of space are made of the same stuff we are.

Those first steps in transforming chemistry to a universal science were largely made by a Bavarian glassmaker who never attended university— Joseph Ritter von Fraunhofer. Fraunhofer was primarily a craftsman. He ran one of the premier and most innovative glass producers, called the Optical Institute, in Bavaria during the early 1800s. The Institute was set in a Benedictine monastery in the Alps, and for his workforce Fraunhofer drew upon the skilled, loyal, and dedicated monks, who had a long history of producing stained glass for many buildings across Europe. To the monks, glassmaking was a way to connect with the divine; light was a sacred and spiritual link to God.[1] Glassmaking was a craft that tangibly pursued the *lux veritatis*, the light of truth.

For Fraunhofer, this served his purpose well. Not only were the monks skilled, they were also quiet and good at keeping secrets. Fraunhofer's glassmaking techniques and recipes would become legendary in his own time. Astronomers, mapmakers, and anyone who sought the best lenses and glass wanted what Fraunhofer was making. He was perhaps the Steve Jobs of his day, running an organization cloaked in secrecy, whose products everyone sought.

Fraunhofer's recipe and process for making flint glass and crown glass set him apart from other craftsmen. Flint glass, made with lead oxides, is dense, while crown glass is light by comparison. (Flint glass derives its name from the lead-rich flint nodules found in the silica in Southern England in the mid-1600s. Crown glass gets its name from the glass-blowing process, where a bubble or "crown" of glass is blown and then formed into a plate.) Flint glass has a high index of refraction, while crown glass has a low index of refraction. (The index of refraction is the ratio of the speed of light in a vacuum to the speed of light in a specific material [e.g., glass] at a specific wavelength. Light moves slower in flint glass and bends more than it does in crown glass.)

Fraunhofer was meticulous and obsessed with producing glass that was clear of bubbles, striations, and streaks. This was hard to do, especially because he used a wood-fired oven and not the much hotter, coal-burning ovens the British were using. Eliminating bubbles meant not stirring the molten glass; however, if the glass was not stirred, then the heavy lead sank to the bottom, creating density differences and causing variations in the refractive properties within a piece of glass. Fixing one problem created another.

To solve the stirring problem, Fraunhofer invented a stirring rod with a wood core encased in clay. The rod was heated, degassed, dipped into molten glass, and repeatedly heated until all indications of trapped gases were gone. The rod was then hooked into a mechanical stirring mechanism that slowly churned the molten glass without introducing a host of new gas bubbles. This "high-tech" stirring process kept the bubbles out and the density uniform.

Along with manufacturing quality glass, Fraunhofer also wanted to know the refractive properties of every kind of glass he could make: How does light bend as it goes through the glass, and how do the various colors bend in different ways? Figuring out the refractive properties of glass was (and frankly is) a very hard problem. Isaac Newton, and many who followed him, had tried and given up. Fraunhofer had no deep scientific curiosity, he simply wanted to provide quality, uniform lenses to whomever needed them. His obsession with his craft drove him to be a great scientist.

Around 1814, Fraunhofer attached one of his carefully crafted prisms to the far end of a telescope. (Technically, it was a theodolite, a telescope mounted on a platform used to measure angles when surveying landscapes.) He then placed a lamp behind a board with a small slit in it; this created a rainbow in much the same way we do today with a glass prism. By using the telescope to look at the spectrum coming out of the prism, Fraunhofer could carefully measure the angles at which different colors emerged from the glass. This allowed him to measure the total dispersion (essentially how wide the rainbow was), as well as the angles for different colors. What Fraunhofer invented was an early version of the spectroscope. (To be fair, Newton had built something similar, but not nearly as capable.)

Fraunhofer soon learned, however, that his spectroscope was limited in its ability to see all the detail within a spectrum. He wanted to measure the refractive, or light-bending, properties of his various glass mixtures; to do that he needed to be able to examine each color in the spectrum very carefully. In addition, he needed to know the exact angles at which the light entered the prism and the angle at which it left the prism.

But there was a very basic and hard problem he needed to solve. Light from the lamp spread out in all directions (just as it does from a light bulb), and even when he put a slit in front of the lamp to limit the light to one particular beam, the light hitting the prism came in at a variety of angles and fanned out into a spectrum that was hard to analyze. He wanted the light coming into his telescope to be parallel so he could better control the path through the optics of the telescope. One way to get nearly parallel light is to move farther away from the light source. Fraunhofer could do that, but then the fan of the spectrum was so large that he could not view the entire range through his telescope. Again, solving one problem created another.

Fraunhofer wanted a straight beam of light with all the colors traveling parallel to each other and hitting the prism at the same exact angle.

If he could figure that out, then he could measure all the colors and all the angles. The solution that Fraunhofer came up with was really quite ingenious and it set the stage for measuring the spectra of a variety of substances.

Instead of using only one lamp, Fraunhofer used six lamps, one for each major color: red, orange, yellow, green, blue, indigo/violet. (I believe Fraunhofer clumped indigo and violet together.) The lamps were arranged in a line and each was placed behind its own slit. The light passing through the six slits was then directed onto one prism. As the light passed through the prism, six rainbows were created. Fraunhofer then adjusted the arrangement such that the six rainbows overlapped in just the right way so as to create one, singular tight beam of a rainbow, with each lamp providing just part of the rainbow. He then passed that rainbow through another prism, which was situated far away (225 meters!), ensuring that the incoming beams of light were nearly parallel. After passing through the second prism, Fraunhofer was able to use his telescope to observe the full spectrum in incredible detail (Figure 3.1). With his clever arrangement Fraunhofer was able to measure the index of refraction for each color of light to six decimal places. He was also now able to see very specific bands, or lines, with the spectrum where the color increased or decreased in brightness.

In his earlier work, Fraunhofer had observed a relatively bright orange line in his measurements of spectra from his lamps, which were fueled by alcohol and contained sodium. Now, with his six-lamp set up he could see that this bright line was really two lines, and he could measure the distance between these lines and their position in the full spectrum. He had discovered a spectroscopic fingerprint for sodium.

Curious to see if specific lines could be found in the spectra of other materials, Fraunhofer experimented with different flames and substances, each time observing patterns in the spectral lines that served as a fingerprint for the substance. Before long, he configured his arrangement so that he could study the spectrum of the ultimate light source—the Sun.

Fraunhofer discovered a vast array of spectral lines (over 550) within the rainbow of the Sun. He noticed that some of the lines in the spectrum of the Sun were similar to some of the lines he had seen by burning

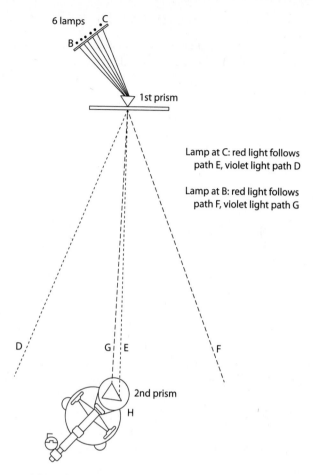

6 lamps

B

C

1st prism

Lamp at C: red light follows
path E, violet light path D

Lamp at B: red light follows
path F, violet light path G

D

G E

F

2nd prism

H

FIGURE 3.1. Splitting the Sun. Fraunhofer's six lamps experiment was the first to dissect light into a detailed spectrum. As light from six lamps (positioned at points B through C) passed through the prism (at point A), the light was spread out. Some of that light reached the prism and lenses at point H, where Fraunhofer could then view and sketch the spectrum. The arrangement of lamps generated a full spectrum at the prism at H. The light from lamp B is spread by prism A into the fan of light shown by lines F and G. The red light from lamp B traveled along line F, while the violet light traveled along line G. Meanwhile, for lamp C, the red light traveled along line E, while the violet light traveled along line D. Fraunhofer could then compare the spectrum at H (with violet to red spanning G to E) to what he saw when he looked at the Sun. The dark lines that he saw in the Sun's spectrum indicated that a more complicated process was occurring than what he saw in the lamps. (Adapted from Jackson, M. W. (2000). Spectrum of belief: Joseph von Fraunhofer and the craft of precision optics. Cambridge, MA: MIT Press)

various materials in his lamps. Were those materials part of the Sun? How could that be? The Sun was a magical burning orb in the celestial sphere. How could it be made of some of the same materials found on Earth? Fraunhofer's spectroscope was the first real window into the chemistry of the cosmos, providing humans with knowledge that could link chemistry on Earth to worlds and wonders beyond. Fraunhofer began the process of making these grand connections, but it would take a few more decades to fully connect the dots between chemistry on Earth and that of the Cosmos.

CHEMISTRY IN THE COSMOS

For all of his technical brilliance and meticulous craftsmanship, Fraunhofer was perhaps too far ahead of his time when it came to the science of chemistry. He saw the patterns in the spectral lines of everything he measured, but the idea of a periodic table of elements was still a few decades away.

By 1815 only 45 chemical elements were known. Although they were known, the concept of an element was still poorly formulated. There was no strong framework yet for atoms with electrons, and nuclei filled with protons and neutrons. John Dalton's *A New System of Chemical Philosophy* was first published in 1808 and paved the way for a detailed atomic theory. But a host of experiments were needed to connect Dalton's "philosophy" to the world around us. At the center of that connection was spectroscopy.

Discovering the link between spectroscopic lines in a rainbow and the true nature of chemistry would fall to a curious pair of Germans in the 1860s. The elder of the pair, Robert Bunsen, would become famous in high school chemistry classrooms for designing a burner with a flame that has little to no color, the so-called Bunsen burner. One important thing about a Bunsen burner is that if you burn something in the flame and it makes a color, the color is due to whatever elements were in that material. The younger of the pair, Gustav Kirchhoff, had already made significant contributions to electricity by the time he partnered with Bunsen and turned his attention to the elements.

Bunsen and Kirchhoff are, for the most part, credited with relating specific lines in a spectrum to individual elements and compounds. Building off Fraunhofer's innovations, Bunsen and Kirchoff arranged their spectroscope so they could simultaneously compare the lines in the spectrum of the Sun with spectra of salts and other materials. In some cases, such as sodium, they found nearly perfect matches. They were the first to make detailed connections between the lines that Fraunhofer had observed in the spectrum of the Sun and the lines they saw in their laboratory spectra.

The pair also figured out why some of the lines Fraunhofer observed were dark. By using a bright background flame as a source, they placed gases of various elements in front of the flame and discovered absorption lines. Elements not only emit light and create bright lines, they also absorb background light, creating dark lines from absorption in spectra.

Although Fraunhofer had arguably done much of the hardest work leading up to the discoveries of Bunsen and Kirchoff, he was limited by the concept of an atom and an element. By the late 1800s, however, "elements" had been organized into an early version of the periodic table by Dmitri Mendeleev. Bunsen and Kirchoff were left to make the cosmic connection: the elements on Earth have a spectroscopic fingerprint, and those fingerprints of the elements could be seen in the spectrum of the Sun.

This was astounding. *The elements in the Sun are the same as the elements found here on Earth.* For the first time in human history, it was understood that chemistry *works* beyond Earth: that is, the elements of Earth are also present in the cosmos. Connecting the chemistry of our Sun to the chemistry of Earth was only the first step in a chemical revolution made possible largely through the study of rainbows.

Understanding spectroscopy, and how light interacts with matter, is central to understanding how we learned about the surface of Europa. Spectroscopy is possible because the wavelengths at which atoms and compounds absorb and emit light is directly related to their structure at the atomic and molecular level. At the atomic level, electrons move up and down energy levels within the atom's electron cloud as they absorb and emit light. At the molecular level, atoms that are connected to each other vibrate in different ways as they absorb and emit light.

In some ways, there is a good analogy to sound waves and how they are absorbed and reflected in different environments. You have probably encountered situations where certain frequencies of sound carry through a room better than others. Perhaps you've adjusted the pitch of your voice in a crowded restaurant or had to lean in toward a friend whose voice seems to get absorbed by the surrounding environment. Some wavelengths of sound bounce off walls and objects in a room, while others get absorbed. This is similar to what happens to light when it passes through, or bounces off, different materials. Light of different wavelengths is preferentially absorbed, transmitted, or reflected off materials depending on how it interacts with the electrons in the atoms, and the bonds that connect atoms into compounds.

So far I've focused on spectroscopy with *visible* wavelength photons (i.e., red to violet). This is because in the early days of chemistry the only real "detector" or "sensor" was the human eye. Fraunhofer, Bunsen, and Kirchoff all had to use their eyes to figure out which colors in the spectrum were brighter and darker than others. Spectroscopy with visible light is very useful, but it's only part of the story.

The infrared region of the spectrum is particularly useful for determining the compounds and molecular structure of a substance. These longer wavelength photons have energies that correspond to the energies of vibration between atoms within molecules. Thus when infrared light is absorbed or emitted, we can tell which molecule (or at least which class of molecules) are responsible for the lines measured in an infrared spectrum. Returning to the analogy with sound, if you turn the base on your stereo up too high, things will start to shake as they absorb and release the energy from those long wavelength sound vibrations. Infrared light does something similar at the molecular scale: molecules shake and vibrate, absorbing and emitting different wavelengths of infrared light.

Interestingly, infrared light was discovered in 1800 by astronomer John Frederick William Herschel, but it was not until the 1940s, when lead sulfide detectors were developed, that infrared spectroscopy could really take off. Herschel cleverly used thermometers to measure the temperature of each color band in the rainbows he made with a prism. He noticed that a thermometer placed a little bit past the red band, where it appeared there was no light, showed a rise in temperature. From this he inferred that there must be invisible "heat" light.

Today we are quite familiar with heat sensors, infrared cameras, and the use of infrared light for everything from remote controls to security equipment. New detectors and sensors have allowed us to greatly exceed our biological sensing capabilities. Although we cannot "see" in the infrared spectrum, we have tools like cameras and spectrometers to extended our capabilities. Those tools, when coupled with telescopes, have allowed us to see water, in its icy form, on the surfaces of worlds like Europa.

A WORLD COVERED WITH ICE

Water's behavior across the electromagnetic spectrum is rather complex for what would appear to be a relatively simple molecule. In the visible wavelength region, water is basically transparent in liquid and solid form. In other words, you can see through it. With the exception of blue light, most wavelengths of visible light pass through water relatively easily or are absorbed by the water molecules. But blue light is scattered around by water, hence the blue color of the ocean and of our skies.

Solid ice is pretty transparent too, but as ice crystals become smaller and smaller, the surfaces of the crystals act like little mirrors, bouncing the light all around and back out to your eye. This is why snow is white: light passes through the crystals but then bounces off the edges like a glistening chandelier, sending all the colors back out, which we see as white.

Moving to longer wavelengths and into the infrared region, water becomes much less transparent. Water does not let infrared photons pass easily through it. The structure of water, and the spacing and strength of the bonds between oxygen and hydrogen, are such that they absorb infrared photons quite well. For this reason, spectroscopy in the infrared is very useful for studying ice and water.

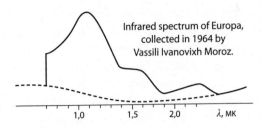

FIGURE 3.2. This spectrum, collected by Russian Vassili Moroz on October 1, 1964, is one of the first detailed spectra ever taken of Europa's surface. The distinct shape of the spectrum, with notches near 1.5 and 2.0 microns (millionths of a meter), is strongly indicative of water ice. (Adapted from Moroz, V. I. (1966). Infrared spectrophotometry of the Moon and the Galilean satellites of Jupiter. *Soviet Astronomy, 9, 999*)

Starting in the late 1950s, Russian astronomer Vassili Moroz began experimenting with an infrared spectrometer he built and adapted for a telescope he was using in Crimea.[2] Early on, Moroz targeted Venus and Mars, but by the mid-1960s he collected the first infrared spectra of Jupiter's large moons.

What he saw on Europa transformed it from a curious speck of light to a world covered in ice. Moroz's spectrum—his rainbow in the infrared—showed two strong absorptions at 1.5 and 2.0 microns (Figure 3.2), creating a staircase-like shape highly characteristic of water ice. The planetary scientist Gerhard Kuiper (of Kuiper belt fame) soon corroborated these results, and by the early 1970s, finer instruments clearly showed that Europa, Ganymede, and Callisto were ice-covered moons. Some 350 years after Galileo discovered these moons, we now had evidence that they were covered in ice.

Toward the end of the 1970s, we finally got a chance to see all that ice up close and corroborate the spectroscopy measurements made from Earth. The *Voyager 1* and *2* spacecraft flew by Jupiter in March and July 1979, respectively. Each flyby, from the inbound approach to the outbound retreat, lasted only a couple of weeks. The actual amount of time each spacecraft spent within Callisto's orbit of Jupiter was about 65 hours. (Recall that Callisto is the outermost of Jupiter's large moons. Ganymede, Europa, and Io are all closer in and have smaller orbits around Jupiter.)

The Voyager flybys were fast and packed with measurements and observations of Jupiter and its moons.

Voyager 1 flew very close by Io, Ganymede, and Callisto but was unable to approach Europa at close range—the nearest it got was almost twice the distance from the Earth to the Moon. The most astonishing discovery from that mission, as described in chapter 2, was about Io. Soon after the images came down, Linda Morabito, a spacecraft navigation engineer, noticed an odd "cloud" above the surface of Io. That cloud, she and the team determined, was a snapshot of a volcanic eruption on Io.

Upon further processing of the image, a dome-like volcanic plume could be seen clearly, and additional images revealed similar plumes. Volcanos were erupting on Io—an astounding discovery, then as now. These images confirmed the amazing prediction of Peale, Cassen, and Reynolds, published a few days prior to the *Voyager 1* flyby. In a matter of days, their idea went from hypothesis to fact.

With *Voyager 1* data still fresh, *Voyager 2* made its flyby and passed within 206,000 kilometers of Europa's surface. The fractured, icy surface of Europa came into full view. The tidal tug that made Io so volcanically active appeared to manifest itself as a maze of criss-crossing cracks in the ice of Europa's surface. Like the shell of a hardboiled egg rolled between your hand and a table, the icy shell of Europa told the story of a world compressed and stretched by tidal flexing.

We now had images—visual proof—that Europa was covered by a shell of water ice. Those images also showed that Europa had very few craters. To a planetary scientist this is a clue indicating the surface is young. Some form of geologic activity had erased any craters made by asteroids, repaving it with fresh ice.

For the first time, the scientific community began to think there might be an ocean beneath the ice.[3] The story of the tides and the energy they might provide, coupled with the images and spectra of a fresh, young, icy surface on Europa, opened up the prospect of liquid water hidden below. If tides create a molten interior of lava that erupts on Io, then perhaps tides can maintain liquid water beneath Europa's icy crust.[4]

But with those images came additional questions. Why did the fractures in the ice look reddish-brown, and why were some regions on

Europa darker red than others? What other compounds were on Europa's surface?

To answer these questions would require another spacecraft. That spacecraft carried the name of Europa's discoverer, *Galileo*.

The *Galileo* spacecraft, which was launched in 1989 and orbited Jupiter from 1995 to 2003, was the first spacecraft to orbit one of the large planets in the outer solar system. The four spacecraft that went before *Galileo—Voyager 1* and *2*, and *Pioneer 10* and *11*—all made quick flybys, sling-shotting around Jupiter as they made their way to more distant destinations. The *Galileo* spacecraft orbited Jupiter, dancing throughout the Jovian system for nearly eight years, all the while making flybys of the moons and sending back stunning images and amazing data (Figure 3.3).

There are many stories about the *Galileo* mission, some of which have been told in other books, and some of which may unfortunately never see the light of day. It was a bold mission that survived political turmoil (funding cuts to NASA in the 1970s and 1980s), national tragedy (the Space Shuttle *Challenger* accident of 1986), and engineering problems of its own (the large antenna on the spacecraft failed to deploy correctly). Here is just one story about the challenges of getting a spectrometer on board the spacecraft to work.

Among the many extraordinary instruments *Galileo* carried was the Near-Infrared Mapping Spectrometer (NIMS), whose purpose was to collect spectra to determine the composition of Jupiter and its many moons. It ultimately proved critical to understanding Europa's chemistry, but it very nearly didn't survive the journey to the Jovian system.

The person in charge of the NIMS, JPL scientist Dr. Bob Carlson (now retired), was and continues to be a friend and mentor of mine. Bob is a humble genius. He has an incredible intuition for how photons behave, which is a very good quality for a spectroscopist.

The NIMS was designed to collect light and create spectra across the wavelength range from 0.7 to 5.2 microns. Once you get into the infrared region of light, beyond about 1 micron in wavelength, it becomes very hard to distinguish the photons you are trying to collect from photons that are due to heat from the instrument itself. To compensate for heat

FIGURE 3.3. Europa's surface, from *Galileo*, December 1997. This grayscale image covers an area 1 mile wide; each pixel represents approximately 20 feet (6 m); everything in white is predominantly water ice, while the dark gray and black regions could be salts or other materials delivered to, or collected on, Europa's surface. In the center is an ice cliff, a few hundred feet high, perhaps created as tidal activity disrupted the terrain, causing uplift. The dark material at the base of the cliff might be from an ocean below. (Photograph from NASA/JPL)

from the instrument, engineers typically figure out ways to cool instruments so that the number of thermal photons are greatly reduced.

Bob and the team designed a passive cooling system on the NIMS, essentially a metal funnel that conducted heat out of the instrument and radiated it into space. It was an elegant, lightweight, no-power solution.

There was another problem though. The *Galileo* spacecraft was launching from Florida in October 1989 as part of the Space Shuttle *Atlantis* payload. Florida is wet, hot, and humid, and all those pesky water molecules in the air love to sneak into the nooks and crannies of a spacecraft.

What this means is that after launch, in the vacuum of space, a spacecraft typically goes through a period of "outgassing," where all the molecules that filled the gaps bounce out and around the spacecraft. Eventually the spacecraft loses this water, but it can take some time. If there happens to be a very cold region on the spacecraft, some of the water molecules might stick to it and start to form a film of ice. There was concern that the cooling funnel on the NIMS might cause ice to get trapped on the lenses and other parts of the instrument.

To prevent ice from forming on the lenses, covers were built for the optics and cooler, and dry nitrogen gas was pumped through the system right up until launch to prevent the humid air from getting in. While this kept the water out, the covers added a bit of additional complexity. The covers had to be blasted off the instrument once the whole system was in space. And this is where things got dicey.

The plan was to fire little explosive devices, called "squibs," to release the covers a few weeks before the spacecraft made a flyby of Venus. In the aerospace world, squibs are very reliable, fancy firecrackers that are used in all sorts of applications to release a valve or to perform a small action.

When the time came to fire the squibs, Bob and his team sent the command and then waited for the temperature of the NIMS to drop, since the cooling system would now be working.

But nothing happened.

They waited, and waited, and waited.

Still, no change.

The team struggled to figure out what had gone wrong. When it comes to complicated missions, there are thousands of things that can go wrong, and for every one thing you can think of, there are dozens of things that you're not thinking of.

Eventually the team concluded that the metal of the cooling funnel might have become slightly distorted due to heating, trapping the cover like a tight lid on a jar. The team sent a command to the spacecraft to turn off a heater in the instrument, and then waited nervously as this miraculous hunk of metal in space implemented their command. The heater shut down, the metal funnel cooled, and the cover popped off. To the relief of everyone on the mission, and Bob in particular, the NIMS was back in business. The spectrometer was now safely on its way to Jupiter, charting a path toward many great discoveries.

RAREFIED RAINBOWS IN THE INFRARED

Roughly forty years after Moroz collected the first spectra of Europa, the *Galileo* spacecraft began orbiting Jupiter and sending back spectra captured from a much closer vantage point. With an instrument so close and able to collect spectra of small features on the surface, new insights into Europa's chemistry became possible. The line of a spectrum is kind of like a shoreline: the more detail you have, the more you can zoom in, and therefore the more you learn about the landscape. Each little nook and cranny of a spectrum tells the tale of some chemical compound absorbing or emitting light. With the *Galileo* spacecraft flying past Europa, the NIMS collected spectra with lots of new details.

First and foremost, the NIMS data corroborated the earlier observations indicating water ice. While important, this was not surprising. The finer features in the spectra, however, revealed new and interesting chemistry. These new spectra catalyzed a long-standing debate about the composition of Europa's surface.

Galileo and *Voyager* images revealed that while Europa's surface is mostly a bright white, consistent with small grains of water ice, many regions also showed a yellowish-reddish-brown color of something that was clearly not water ice. This material appeared to map onto cracks and other features on Europa's icy shell that might connect to an ocean below. For this reason, some scientists hypothesized that the yellowish-reddish-brown materials could be salts or other compounds dissolved in the ocean and erupted onto the surface.

To further investigate the composition of this material, the NIMS team collected spectra from a variety of places on Europa's surface. The data revealed that this dark "non-ice" material is dominated by sulfate, which is composed of one sulfur atom connected to four oxygen atoms.

Sulfate is found in all sorts of places on Earth, and you may have even taken a hot, relaxing bath in it. Epsom salts, which you can buy at a pharmacy and pour into your bathtub, are crystals of magnesium sulfate. Magnesium joining with sulfate to form Epsom salts $(MgSO_4)$ is one example of where you can find sulfate on Earth.

Another, much less soothing example is sulfuric acid. Sulfuric acid is the combination of sulfate with two hydrogen atoms (H_2SO_4), yielding an incredibly toxic and corrosive compound.

Curiously, the two leading candidates for matching the NIMS's spectra of the material on Europa were (and continue to be) magnesium sulfate and sulfuric acid. The difference between these two materials and their implications is tremendous. On the one hand, if the material is Epsom salts, then that indicates the material on the surface is likely from a salty ocean below, and that is a very exciting discovery. Magnesium sulfate is found in our ocean on Earth, and it is found in all sorts of hot springs and other environments where water and rocks flow and mix together. To see Epsom salts on Europa's icy surface would imply upwelling of salty ocean water which then freezes and traps the salt on the surface. The ice shell would be a window into the chemistry of an ocean below.

On the other hand, if the material on Europa's surface is sulfuric acid, then the story of a salty ocean below no longer hangs together by merit of the spectra. Sulfuric acid would indicate that sulfur ejected from volcanoes on Io travels through space to Europa's surface, where it bombards the surface ice and is processed into sulfuric acid (sulfur mixed with ice). It is still an exciting discovery—sulfuric acid on Europa made from sulfur ejected from volcanoes on Io! Wow! But it's not evidence for a subsurface ocean.

Unfortunately, the spectra from the *Galileo* mission did not have quite enough detail to help definitively determine which hypothesis is correct. Salts are not strictly required to explain most of the data, but they do potentially match up better with some of the spectra collected from cracks and other regions where one could envision material from an ocean coming up to the surface. My colleagues Brad Dalton and Jim Shirley, both at JPL, have carefully examined this issue and concluded that sulfate salts like Epsom salts are needed to explain some of the NIMS's spectra.

But Bob Carlson, who knows every detail of the NIMS, has strong reservations about overinterpreting the data; sulfuric acid provides a very good match for the available data, and there is no need to mix in salts to get a better fit. The extraordinary claim of oceanic salts is not sufficiently supported by the available spectroscopic evidence. The salt-versus-sulfuric-acid debate has gone back and forth for over two decades, with no real end in sight, until we can get another spacecraft with new instruments back out there.

There has been one great improvement, however. In the decades since the NIMS was built, telescopes here on Earth, and those in space—such as the Hubble Space Telescope—have gotten much better, as have the spectrometers used with those telescopes. Recent work using these tools has turned up some good evidence for salts on Europa's surface.

I and my colleague, California Institute of Technology Professor Mike Brown (also known as the man who killed Pluto), along with several of his students, used one of the large telescopes at the Keck Observatory in Hawaii to capture new spectra of Europa's surface. The Keck telescopes are an incredible testament to how far spectroscopy has come since the construction of the *Galileo* spacecraft and the NIMS in the late 1970s. With the Keck's ten-meter-diameter mirror (approximately 30 feet), adaptive optics, and spectrometer, we were able to capture spectra with forty times as much detail in the spectra relative to those collected using the NIMS. Our spectra include never-before-seen lines, and we think some of these new lines are due to salts—salts that would almost certainly have to come from an ocean below.

Following on that work, Mike's brilliant graduate student Samantha Trumbo led an effort to gather some spectra of Europa using the Hubble

Space Telescope. Our goal was to see if Europa had a specific set of absorptions that we had seen in the lab after irradiating sodium chloride (NaCl), which is the most abundant salt in Earth's ocean. Sodium chloride is table salt and is typically white in color. However, after being bombarded by energetic electrons—which is what happens on Europa's surface—the salt turns a yellowish-brown color and develops a very distinct spectroscopic fingerprint. Sure enough, when Samantha got the Hubble data and dug into the spectra, she found a match to our lab results. It's hard to explain the data any other way, and thus I think we now know that there is sodium chloride on Europa's surface. That salt comes from the ocean below. I like to think that perhaps, just maybe, mixed in with those salts are relics of life within the ocean.

———

The advent of spectroscopy and other technologies from the mid-1800s to the mid-1900s brought about a tremendous transition in astronomy: instead of making telescopic observations with their eyes only, astronomers were now able to collect images—and ultimately spectra—of their targets.

For about 350 years, Europa was simply a bright spot of light circling Jupiter. Although astronomers had carefully watched the moons of Jupiter since Galileo's great discovery in 1610, they had made no progress in figuring out the composition of the moons. With the powerful tool of infrared spectroscopy, however, Moroz, Kuiper, and others unveiled the first piece of the puzzle in figuring out that Europa has an ocean. Spectroscopy allowed scientists to determine that Europa's surface is made of water ice. This rainbow connection was the first big step toward the discovery of an alien ocean.

CHAPTER 4

BABYSITTING A SPACECRAFT

A babysitter watches a child and may, if he or she is lucky, have a sense of what they are in for. The child may run around the house, bumping into furniture, playing with toys, and making a mess with their food. However well-prepared the babysitter is, the child always behaves a bit differently than expected.

Spacecraft are the same way. First, they need a babysitter. Second, they don't always behave the way you thought they would, and sometimes that is a really good thing.

For the first piece of the ocean worlds puzzle, spectroscopic observations revealed water ice on Europa's surface. But those observations were only skin deep—infrared spectroscopy sensed only approximately the top 100 microns of Europa's surface. Those measurements were unable to penetrate the ice shell to tell us about what lay beneath. Perhaps Europa was covered with just a thin veneer of ice, beneath which was a world of rocks.

Getting below the surface required babysitting a spacecraft. By carefully tracking (or babysitting) the *Galileo* spacecraft with large antennas here on Earth, scientists could make gravity measurements of Europa. Gravity measurements are the second piece of the puzzle. Gravity measurements helped reveal what lay within Europa. How is that possible and what does it even mean to make a 'gravity measurement'?

Gravity is a measure of the shape of space. The shape of space is determined by mass. As an object travels through space, its path, or trajectory, is determined by the shape of space. Just like a golf ball rolling on a curvy golf green, a spacecraft moves through the solar system over a landscape where the curvy valleys are created by the Sun, planets, and moons.

Another analogy often used for gravity and the shape of space is that of a bowling ball on a mattress; massive objects like stars, planets, and moons, create gravity "wells" like the depressions made in the mattress. Gravity wells cause objects to fall in toward them, like a marble rolling into the depression made by a bowling ball on a mattress. Massive objects in our universe—like galaxies, black holes, stars, and planets—warp the shape of space and time, and our experience of gravity is a result of these distortions in space. Our understanding of gravity deforming space and time goes back to Einstein and his general theory of relativity, which made the link between the force of gravity and the shape of space. Incredibly, over a hundred years after Einstein's great insight, scientists are now measuring gravity waves and taking images of black holes, both of which demonstrate the connection between mass, gravity, and the fabric of space itself.

But our story, and the mystery of Europa's ocean, takes place on a much smaller scale. In our own local neighborhood of the solar system, the Sun creates one big gravity well in space-time; the planets create smaller depressions that orbit the gravity well of the Sun; and finally moons of those planets create even tinier wells orbiting the depressions made by planets. We could keep on going with this down to the finest scale: even asteroids and comets make tiny little gravity wells, adding to the complex texture of space-time (Figure 4.1). For our story, we follow the path of the tiny *Galileo* spacecraft as it sails up and down a host of gravity wells before eventually reaching the gravity well of Europa.

THE BABY

The "baby" in this babysitting analogy is the *Galileo* spacecraft. Imagine you were able to hitchhike on the *Galileo* spacecraft. Your path is actually a journey through the various gravity wells that you encounter along the way. The first gravity well you have to escape is the Earth's. *Galileo*

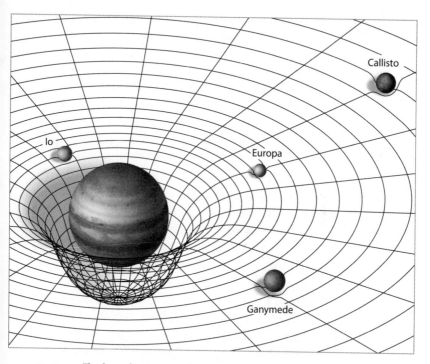

FIGURE 4.1. The shape of space in our solar system and the gravity wells from various objects. Massive objects, like stars, planets, and moons, distort the shape of space itself, creating gravity wells that can be envisioned as divots in the fabric of space. Throughout our solar system, the Sun creates one big, central gravity well, around which the smaller depressions created by the planets and moons orbit.

was launched on the Space Shuttle *Atlantis* and deployed from the payload bay, after which it fired another engine to send it beyond Earth.

As soon as you rise up and out of the Earth's gravity well, you realize you are in a much bigger gravity well created by the Sun. If you want to travel farther out in the solar system, it's going to take a lot of propulsion to lift you up and out of that well. You realize you have only climbed a small hill during your escape from Earth.

The well of the Sun is so deep that you cannot travel straight to Jupiter because you don't have enough rocket fuel. Instead, clever engineers have figured out that you can use the gravity well of the Sun, Venus, and the Earth to your advantage. Instead of heading outward, you and the *Galileo* spacecraft head inward, toward Venus. As you do, you accelerate because you are heading downhill, gravitational speaking.

In addition, you find yourself now chasing Venus. Its moving fast in its orbit around the Sun and now it's pulling on you. The well of Venus draws you toward it, forcing you to accelerate and giving you more energy that you can use to slingshot farther out in the solar system. The path for *Galileo* involved three slingshots, or gravity assists: one from Venus and two from Earth.

Once you reach Jupiter, though, you begin to fall into the gravity well of Jupiter. The influence of the Sun's gravity takes a back seat to Jupiter's, and you become trapped in Jupiter's gravity well. As you orbit Jupiter, you make flybys of the large moons—Io, Europa, Ganymede, and Callisto—and with each flyby you feel a little gravitational bump in the road. Although these moons are small compared to Jupiter, they still distort the shape of space.

While these moons are large compared to other moons in our solar system, they make the tiniest of divots in the fabric of gravity that encircles Jupiter. These tiny divots orbit Jupiter; if you fly close enough to the moons, you can feel the influence of the small gravity wells created by these moons. It's a bit of a Russian doll effect: the gravity wells of the moons are impressions within the large well of Jupiter, and Jupiter's gravity well is contained within the larger well of our Sun (which is the tiniest pin-prick in the gravity structure of our galaxy, and in the universe of hundreds of billions of galaxies).

Zooming in on the gravity well of Europa, you notice that it's slightly distorted and elongated along the direction between it and Jupiter. This is because Europa is slightly stretched along the line that connects it to Jupiter. As we learned in earlier chapters, Europa is tidally locked to Jupiter, which means that it always points the same side toward Jupiter. Europa is also a little squashed, or oblate, because of its own rotation on its axis. This oblateness is pretty common for spinning planets and moons. Even the Earth is a little squashed.

The mass and oblateness of Europa affect the shape of its gravity well. In addition, the *distribution* of mass within Europa also affects the well.

In other words, if there are layers or regions within Europa that have different densities, then those variations will affect the shape of the gravity well too. This is critically important, and unfortunately it's also where the analogy to a bowling ball on a mattress breaks down. If I give you two bowling balls of the same size and weight, they will make the same depression in the mattress, even if their interiors are different. For example, if one bowling ball has a heavy core of iron surrounded by very light plastic, while the other is made of only one material of uniform density, the two depressions in the mattress will still look the same. When it comes to gravity, however, this is not the case. The shape of a gravity well is affected by the variations in the density of the materials within a world.

From your point of view, riding along with the *Galileo* spacecraft, this means that as you fly past Europa, you can feel small variations in the gravitational tug of Europa. Those variations cause the spacecraft to speed up or slow down ever so slightly as it passes through the gravity well. Each little change that the spacecraft experiences can be related back to information about the interior structure of Europa.

To reiterate and emphasize this pretty amazing point, the mass, density, and interior structure of a world, such as Europa, determines the fine details of its gravity well, which in turn affects the motion of a spacecraft as it flies by that world. Therefore, if you can very carefully track signals coming from that spacecraft (e.g., *Galileo*), then you can reconstruct its path and in turn, figure out some of the interior structure of that world (e.g., Europa).

THE BABYSITTER

Carefully tracking a spacecraft's position and velocity is exactly what brilliant engineers and scientists have been able to do for many different missions. They can babysit a spacecraft precisely by using the Deep Space Network (DSN), which is a set of three large radio telescopes positioned around the Earth to receive signals back from spacecraft. The radio dishes are 70 meters in diameter, and they are located 120° longitude apart from each other so that at least one dish will always be able to receive and transmit signals to distant spacecraft. The DSN locations are Barstow, California; Madrid, Spain; and Canberra, Australia.

When a spacecraft transmits data back to the DSN stations, that signal can be used to define the position and velocity of the spacecraft. The spacecraft's position is further understood by knowing when the signal left the spacecraft. This is often provided by an onboard clock, which tags the timing of the data transmission.

These "clocks" often take the form of what is called an ultra-stable oscillator (USO), which is measured before launch and provides a reliable timer for everything that happens throughout the mission. This is critical. Europa is, at its closest, about 390 million miles (630 million km) from the Earth. A signal from a spacecraft flying by Europa takes 36 to 53 minutes to reach Earth. A synchronized onboard clock allows engineers to better determine when and where the data transmission occurred.

Determining the velocity of the spacecraft, and any changes in the velocity that occur as a result of the precise shape of the gravity well, is made possible by careful tracking of the wavelength of the signal on which the data is transmitted. The wavelength of light over which the *Galileo* spacecraft and the Earth (via the DSN) communicated was 13 centimeters, or 2.3 GHz (radio waves often referred to as part of the S-band).

When a spacecraft flies by a world like Europa, the structure of the gravity well causes it to accelerate (on approach) or decelerate (as its finishes its flyby) depending on the mass distribution and shape of the gravity well. These slight variations in acceleration change the spacecraft's velocity; and if the spacecraft is transmitting a signal, that signal is ever-so-slightly stretched or squeezed due to the velocity change. This is the Doppler shift of light, and it pops up in all sorts of useful places in physics and astronomy.

In the case of the *Galileo* mission, a team of scientists and engineers, led by Dr. John Anderson of JPL, measured the Doppler shift of the signal transmitted each time the spacecraft flew by Europa. Not all flybys were useful for this kind of measurement, and thus it was a long and meticulous process over the course of the entire *Galileo* mission. Stunningly, Anderson and his team were able to measure the velocity changes of the spacecraft down to a few millimeters per second. Once those detailed gravity measurements were made, it was off to the races, mathematically speaking. Using the gravity measurements, along with some basic parameters about Europa's size and average density, the team was able to figure out Europa's moment of inertia. Figuring out the moment of inertia was a big step toward figuring out the internal structure of Europa.

A MOMENT FOR EUROPA

The "moment of inertia" of an object sounds much more mysterious than it really is. It simply refers to the inertia—or resistance to change in motion—of a rotating object. It applies to everything from bike wheels to spinning moons.

The term "moment" refers to a force applied at some distance from an axis of rotation. Think of a lever—occasionally you'll hear a lever arm referred to as a "moment arm." The moment of inertia is a similar idea, but it incorporates all of the different masses that are at different points from the center of a rotating object. The term "moment arm" could be used to describe the effort needed to pull on a wrench to release a rusty bolt. For example, "This wrench is too short; I need a longer moment arm to get this bolt loose. Do you have a longer wrench?" The term "moment of inertia," meanwhile, might be used to describe the effort needed to slow down or speed up the rotation of an object. For example, "Don't try to slow down that merry-go-round full of kids; it has a huge moment of inertia and you'll get hurt!"

The moment of inertia for a rotating world like Europa is the sum of all the different moments of inertia for the different layers within Europa. Think of Europa as an onion, where each layer of the onion is made of a different material. Each layer has its own moment of inertia, which is defined by the mass (M) of that layer of material and the radial distance (R) of that layer from the center of Europa.

The mathematical expression for the moment of inertia of a spherical *shell* of material (i.e., one layer in that onion, or a layer within Europa) is $0.66 \times MR^2$. For comparison, if the same amount of mass, M, is distributed into a *full sphere* of material with the same radius, R, then the moment of inertia is $0.4 \times MR^2$. What is important to notice here is that "MR^2" is the same in both cases, but the numbers out front, 0.66 for the shell and 0.4 for the sphere, are different. The shell of material has a larger moment of inertia than the sphere. This should make sense intuitively: a heavy tire that is spinning is harder to slow down than a spinning disk of the same mass because more of the mass is closer to the center, and thus the moment of inertia for the spinning disk is smaller than it is for the spinning tire.

So what is the situation at Europa? After all those careful measurements using the DSN and the *Galileo* spacecraft, the gravity measurements revealed that Europa's moment of inertia is $0.346 \times MR^2$, where M and R refer to Europa's total mass and its radius. For our purposes, all that matters is the number out front: 0.346, which is smaller than the number out front for a spherical shell. What does that mean? It means that Europa is *not* a massive hollow shell. That's no surprise; physically that wouldn't make sense. But it was plausible for Europa to be a solid sphere with the same material throughout, so a value of 0.4 would not have been impossible. That Europa's number is significantly lower than this means that the density of Europa must not be uniform throughout. In other words, Europa must have at least a few layers of different densities.

To understand why a value less than 0.4 means there must be layers within Europa, consider that in the expression above, M and R refer to Europa's *total* mass and radius. If I build a world like Europa with shells of different material, each of those shells will have their own mass (m) and radius (r). The moment of inertia number for each of those shells will still be 0.66, as per the description above. But now if I combine all of those shells, like layers of an onion, the total would look something like: $0.66m_a r_a^2 + 0.66m_b r_b^2 + 0.66m_c r_c^2 + \ldots$, where the subscripts a, b, and c refer to the different layers. With the exception of the outermost layer, the radii of the inner shells must be smaller than the total radius, R. For example, r_b might be half the full radius, which would mean $r_b = R/2$. If we now plug in $R/2$ for r_b in the expression above, we see that ½ becomes ¼ after it's squared; and then it divides into 0.66, and the new term is $0.16m_b R$. In other words, since the radius and masses for each of the shells can only be a fraction of the total radius and mass, the total moment of inertia ends up being smaller than that of a uniform solid sphere, with a value of 0.4. Once you measure the total moment of inertia, you can use these equations to construct different models for the interior structure of a world.

Knowing that Europa's moment of inertia is 0.346, we can then construct and fine-tune models for different layers within Europa.

First, we can ask if a two-layer model fits the moment of inertia for Europa. In short, it does not—or at least it's relatively hard to get a value of 0.346. A two-layer moon would consist of a core and a rocky mantle, both of which are relatively high density. Even for a wide range of assumptions about the density and radius of the core and mantle, you still end up with a moment of inertia value that is larger than 0.346. This makes intuitive sense, as a two-layer model with two materials that are not that different in density will start to approach the value of a solid sphere.

Adding a third layer to the model for Europa, and specifically a layer of low-density material, starts to fit the moment of inertia well. When a layer of material with a density of about 1 g/cm^3 is placed over a rocky layer with a density of about 3 g/cm^3, and those layers both wrap around a solid iron-rich core, then you can get a moment of inertia of $0.346 \times MR^2$.

What could be that outer layer of material that is 1 g/cm^3? In either ice or liquid form, water has a density of about 1 g/cm^3 and is the best explanation for that low-density outer layer. The gravity measurements were not sensitive enough to distinguish between the density of liquid water and ice.

In summary, a three-layer model of Europa with an iron-rich core of about 600 km in radius, a shell of mantle rocks about 800 km in outer radius, and an outer layer of water about 100–200 km thick fits the 0.346 value well (Figure 4.2).

By carefully tracking the *Galileo* spacecraft—babysitting it ever so closely and measuring every slight change in velocity as it flew by Europa—scientists were able to determine the moment of inertia of Europa and ultimately set some constraints on the interior structure and conclude that Europa has a layer of water about 80–170 km thick.

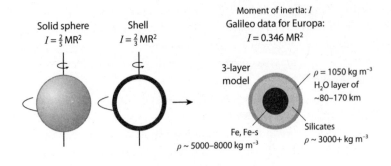

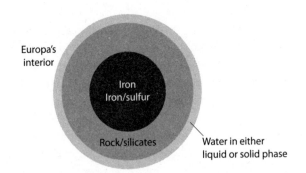

FIGURE 4.2. The moment of inertia of Europa indicates a dense iron core, surrounded by a layer of less dense rocky material, overlaid by a water layer 80–170 km thick in either the solid (ice) or liquid (ocean) phase. Gravity measurements were not sufficient to distinguish between the density of liquid water (about 1.00 g/cm³) and ice (about 0.920 g/cm³).

By babysitting the *Galileo* spacecraft, the second key piece of the puzzle fell into place. Europa not only has water ice on its surface but a water layer—in either ice or liquid form, extending to a depth of over 100 km. Determining that some of this layer is in liquid form—i.e., an ocean of water trapped beneath an icy shell—required one more step.

For that, we must pass through airport security.

CHAPTER 5

HOW I LEARNED TO
LOVE AIRPORT SECURITY

There's a small chance that at some point between 2000 and 2005 I may have been responsible for slowing down your security line at the airport. If so, I apologize. It was all in the name of science.

During that period, I had a habit of testing the limits of the metal detectors. I was not doing this for any sort of security reasons—I had no sharp or malicious objects in my pockets. I wasn't even trying to see how good our security screening was. Instead, I would occasionally walk through the metal detector carrying a small bottle of salt water in my pocket to see if it could set off the alarm. Salt is a good conductor of electricity, and I wanted to see how much salt was needed to set off the alarm.

You see, when *Galileo* flew by Europa in the late 1990s and early 2000s, it was carrying an instrument similar to the sensors in an airport metal detector—a magnetometer (essentially a fancy compass that can detect changes in the strength and direction of magnetic fields). when the *Galileo* magnetometer came close to Europa, the "alarm" went off. In my own simple way, by experimenting with airport security, I was replicating the situation. The physics of airport security is at the heart of the third and final piece of the puzzle leading to the discovery of an ocean on Europa.

The underlying physics is basically the same, whether used to detect dangerous metal objects or to reveal a liquid water ocean on a distant world. In the case of airport security, coils of wire in the side of the door-way that you walk through create a pulsating magnetic field. As you walk through, the rapidly changing field creates electric currents in any con-ductors in your pockets (such as coins). The electric currents in the coins then give rise to a small magnetic field around the coins.

This is what always happens with electricity and magnetism: magnetic fields create electric currents, and electric currents create magnetic fields—they are the yin and yang of each other. The field or current that is created is typically referred to as an "induced" field or current because it was caused by a strong external source, such as the field in the security doorway. Critically, the induced magnetic field of the coins in your pocket can be detected by sensors in the doorway. If an induced field is detected, the alarm goes off. You then get pulled aside and are subject to the pat-down.

Well, when it comes to Europa, the alarm went off. Europa got the pat-down.

In this analogy, Europa is like you walking through the doorway of the metal detector. Jupiter, with its large, changing magnetic field, plays the role of the changing field created by the coils of wire in the metal detector. Lastly, the magnetometer on board *Galileo* is akin to the sensors inside the doorway that detect induced fields and trigger the alarm. Mother nature created her own little alarm system for us, and Europa's ocean set off the alarm.

Throughout the rest of this chapter, we explore the physics behind air-port security and Europa's ocean. But first, we start with a bit of history about how these magnetic field measurements were made and the chal-lenges faced to get the data. We then dive further into this final piece of the puzzle and see exactly how measurements of a magnetic field revealed a global ocean on a distant world.

A MAGNETIC MYSTERY

The detective work revealing the induced field in Europa's ocean was led by Professor Margaret Kivelson at the University of California Los Angeles (UCLA). Professor Kivelson did not aspire to discover an ocean beyond Earth. She told me that, as a woman studying physics in the 1950s, she simply had her heart set on getting a job as a professor at a university.

Although she would make her name in the realm of space physics and planetary science, Kivelson got her start deep in the weeds of quantum mechanics. She began her career with electrons, one of the smallest "things" in our solar system, and went on to make pioneering discoveries of one of the largest "things" in our solar system: Jupiter's magnetic field.

Kivelson was one of 13 students of Nobel Prize–winning physicist Julian Schwinger, and was his only female student. She recalls him fondly as a talented physicist who carefully protected his time. He would spend one afternoon a week meeting with all 13 students. Therefore, being independent and self-motivated was a necessity, and Kivelson developed skills that would help her persevere in the decades to come.

Once she finished her PhD at Harvard University, Kivelson moved from Cambridge, Massachusetts, to the sunny coast of California. The RAND Corporation, based in Malibu, hired her as a scientist to conduct research on plasma physics.

Although she enjoyed the job, she wanted to be a professor. She tried to get a job at UCLA, but it was not until Professor Willard Libby, who invented the technique of carbon-14 dating, had an opening for a supervisor that she got her chance. Libby needed someone to manage his cadre of students, and he saw from Kivelson's application that her skills overlapped with some of the projects his students were working on. Kivelson got the job and left RAND.

One of Libby's students was researching Jupiter, and over the two years Kivelson worked for Libby, she developed an expertise on Jupiter's magnetic field and the interaction of atomic particles, like electrons, with that magnetic field.

Having proven her skills, she was recruited by Tom Farley, another professor at UCLA's Institute of Geophysics, to help with satellites used to study Earth's magnetic field. From there she went through a series of promotions and lateral moves and was then put in charge of a group of scientists tasked with analyzing the data from NASA's twin *Pioneer 10* and *11* spacecraft. The year was 1974. Seventeen years after completing her PhD, Kivelson was finally in a position worthy of her brain.

But she wasn't done yet. She still had her heart set on being a professor. Although she was at the prestigious Institute of Geophysics, she was still research staff. Lacking an opportunity, she made one: she offered to teach for free.

Not surprisingly, Kivelson was a great teacher. Meet her for a minute and you will sense her patience and ability to communicate difficult concepts. Eventually, the university saw fit to pay her for her teaching. After 13 years at UCLA, she finally got the tenured professorial appointment she so desired. She was now Professor Margaret Kivelson.

Meanwhile, NASA had moved on from the *Pioneer* spacecraft. The *Voyager* missions were under construction, and the *Viking* spacecraft were getting ready for launch. In 1975, NASA released a call for proposals to build instruments that would go on the Jupiter orbiter probe—the spacecraft that would eventually be anointed as the *Galileo* mission.

The geophysics team at UCLA wanted to submit a proposal to build a magnetometer to go on this new mission. Although they had a deep bench of experience with magnetometers in general, no one knew much about Jupiter's magnetic field. Except for Kivelson. Her path from studying quantum mechanics to Jupiter had now placed her in a unique position to lead this team.

Unfortunately, their first try failed.

Failure, however, can be good. I myself have failed many times with instrument, mission, and research proposals to NASA. To turn a classic phrase on its head, *failure has to be an option* when you are trying to do new things and push the frontier. Fail early and often, and learn from your failures.

Kivelson and her team were quick studies in the art of failure. Not long after they failed, they won. It turns out that the mission to Jupiter included two components: the orbiter and a probe. The probe would dive into Jupiter's atmosphere and send back a small, but very useful, dataset as it plummeted into Jupiter. The orbiter would circle Jupiter many times and fly by its moons, sending back data from everything of interest in the Jupiter system.

The first NASA request for instruments was for instruments on the probe. While the concept of the probe did not include a magnetometer, Kivelson and her team proposed one anyway. Perhaps not surprisingly, their proposal was not selected.

Although she had failed, Kivelson told me she learned a lot from that process. Therefore, when the call for instruments on the orbiter came out (soon after the call for the probe instruments), Kivelson and her team were ready. They wrote their proposal, submitted it to NASA, and won.

Kivelson would lead the team to study Jupiter's magnetic field on the first spacecraft ever to orbit a giant planet. An exciting, and daunting, prospect.

In 1977 when Kivelson attended the first full meeting of the *Galileo* team, she was one of only two women on the entire team. (The other woman, Margaret Hanner, was helping develop an instrument to study the dust particles around Jupiter.) This mission, Kivelson's instrument, and she as a woman in the field, were all simultaneously on a pioneering voyage.

BIT BY BIT

For all the effort Kivelson and her team had put into building their magnetometer, any instrument is only as good as the data it can send back from space. And sending data back requires a good antenna.

In chapter 3 I mentioned that the *Galileo* mission faced many challenges. One of the biggest challenges was that its primary, high-gain antenna failed to deploy correctly. Without the high-gain antenna working, the flow of data coming back from the instruments slowed to a trickle. Kivelson and her team were forced to rethink how they were going to do their science. In the end, they made some changes that proved to be fortuitous, and they managed to get a lot of the most important bits of data back to Earth.

In today's age of high-speed Internet and terabyte hard drives, it's easy to forget how hard some of these basic functions were for spacecraft designed and built in the 1970s and 1980s. Thankfully for the *Galileo* mission, some clever engineering overcame what could have been a crippling failure.

On April 11, 1991, the computer inside the *Galileo* spacecraft sent commands to motors that should have opened its large umbrella-like antenna. This 16-foot-wide, high-gain antenna would make it possible for the spacecraft to send images and data back at a rate of about 134,400 bits per second, which is equal to 0.134 megabits per second (Mbps). While today we enjoy typical Internet speeds of about 100 Mbps, in the early days of the Internet, when dialing-up through your phone line, data rates could get to 0.056 Mbps on a good connection. The high-gain antenna would have allowed the *Galileo* spacecraft to send data back at about twice the speed of dial-up. Without the high-gain antenna, using only the two small, low-gain backup antennas, the spacecraft would be limited to about 10 bits per second, which is roughly 10,000 times lower than the high-gain antenna.

The commands to open the antenna were sent on that day in the spring of 1991, and engineers back on Earth waited for the results. When the data arrived, it indicated a bad situation: the antenna did not fully deploy. Only a few "ribs" of the umbrella structure had released and extended outward from the closed position.

Teams on the ground worked tirelessly to diagnose the problem and test ideas with an identical version of the antenna. Nothing worked. For five years, between 1991 and 1996, as the spacecraft made its journey to Jupiter, the engineering team tried everything from hammering the motors on and off to generate maximum force on the ribs, to heating parts of the spacecraft with the Sun, to trying to deploy the antennas during periods of maximum thrust. Still, nothing worked.

In the end, the team concluded that the ribs had lost lubricant, and the pins at the ends of the ribs had essentially etched into their sockets, preventing them from springing out into an open umbrella shape. As best they could tell, three of the ribs had successfully deployed, but fifteen were still stuck. The etching problem almost certainly happened as a result of transporting the antenna from Florida to California and back during the years long delay of the spacecraft launch.

Although the team now understood the problem, there was no way to fix it. Through a host of clever upgrades to the onboard software, and by increasing the number and quality of receiving antennas on Earth, the team managed to get the data rate of the backup antennas up to 160 bps. It was a factor of 10 improvement, but still a thousand times worse than if the high-gain antenna had worked.

And the fun did not stop there.

A low data rate might not be a big problem if you can store a lot of data on your spacecraft and then trickle it back over a long period of time. While orbiting Jupiter, the *Galileo* spacecraft would have a lot of time when it was not flying close to anything and could unload stored data.

Remember, however, that the *Galileo* spacecraft was designed and built with 1970s technology. There were no flash drives or other compact hard drives. The technology was reel-to-reel magnetic tape, rolling back and forth. Think eight-track tapes and old cassette tapes.

Galileo's tape recorder was, in fact, a fancy version of an old cassette tape recorder. Cassette tapes from the 1970s, 1980s, and 1990s, were four-track magnetic tapes, which meant that the magnetic tape had four lanes, or tracks, and the magnetic head in the tape recorder would position itself over two of those tracks at a time. Of those two tracks, when music was playing, one track was for the left speaker and one for the right. When the tape finished, you would remove the cassette, flip it over, put it back in, and press play. In so doing, you now gave the magnetic head access to the two tracks on the other side of the tape, one for the left and one for the right speaker.

On *Galileo*, the four-track tape ran back and forth from reel-to-reel as data from the instruments were recorded and then played back for transmission to Earth. The length of tape was sizable, approximately 1,800 feet, and it could store nearly 900 Mb of data. But just like those old tape recorders that would occasionally eat your tape, the *Galileo* recorder had a number of problems of its own.

In October 1995, as the spacecraft was fast approaching Jupiter, the tape recorder rewound the tape fully, but then kept on trying to rewind more. For over 15 hours, the rewinding wheel tugged on the end of the tape.[1] It is amazing that the tape did not tear or get pulled off the reel.

The problem with the rewind was fixed in software, but it was decided that the tape itself could be damaged and at risk of being torn completely. To mitigate against that, the team "buried" the bad part of the tape under 25 wraps around the reel. That left just 305 feet of tape for recording data. Instead of 900 Mb of data, the team now had just over 150 Mb. To put that into perspective, that is nearly 20 Megabytes (MB) of data, or barely enough for ten pictures with your phone camera. The tape recorder worked, but the team would be on a much tighter diet.

Each of the onboard instruments was given a strict data ration. The limitation of the low-gain antenna and the tape recorder meant that all the teams would now make big sacrifices in their data volume. For Kivelson and the magnetometer team, that meant they could not measure the magnetic field three times per second, as they had originally planned.

Thankfully, the magnetometer had its own computer with 4,800 bytes of storage, and a clever engineer by the name of Joe Means had written software that allowed the team to average measurements over arbitrary time intervals. As a result, data could be stored and averaged locally by the instrument, converting many bits of data into just a few bits for the main tape recorder.

Another limiting factor was the spin of the spacecraft. The *Galileo* spacecraft was designed such that part of the spacecraft spun, while the other part was "de-spun" (kept from spinning by special reaction wheels). Spinning of a spacecraft is a common way to stabilize it, a bit like the way a spinning top will stay pointed upright. The "de-spun" part of *Galileo* carried the camera and other instruments that were better off not spinning. Meanwhile, the spinning part carried the magnetometer and other instruments that measured the space environment.

Positioned at the end of a 36-foot boom, the magnetometer whipped around in a circle once every 20 seconds. Ideally, Kivelson wanted to make many measurements per rotation, thus enabling her to understand Jupiter's magnetic field in fine detail and also remove any magnetic field artifacts, i.e., noise, from the spacecraft itself.

With the new rationing in place, however, Kivelson would not be able to make all those measurements. For non-critical flybys, she resigned herself to one measurement every 20 seconds, thinking that at least she would get the same relative position on the spacecraft every time. But the *Galileo* project manager, John Casani, told Kivelson that the best he could do, given all the other instrument and spacecraft constraints, was one measurement every 23 seconds, slightly longer than the rotation period.

Although it was frustrating at first, this setback proved a boon as the mission went on. By having a measurement period slightly longer than the rotation period of the spin stage, Kivelson and her team were able to tease out the artifacts from the spacecraft in greater detail, which ultimately improved the quality of the measurements of Jupiter's magnetic field.

Over the mission's lifetime, the magnetometer generated a map of Jupiter's magnetic field and helped reveal it as one of the largest "structures" in our solar system. You may have looked up and seen Jupiter in the night sky, appearing as a very bright point of light. If you could see its magnetic field, you would see a pulsating onion of magnetic field lines about five times larger than our full Moon. Jupiter and its magnetic field, which is tilted by about 10 degrees from Jupiter's axis of rotation, spin once every 10 hours.

Deep within that huge field are the four moons that Galileo himself discovered. Each moon making its way around Jupiter, passing through the field like travelers at the airport passing through the security line.

EUREKA?

Movies often portray science as a process punctuated by big "Eureka!" moments. Characters struggle to figure out a problem, and in a dramatic montage, the music plays as they toss and turn for days, weeks, years. Then suddenly, like a bolt of lightning, they have an insight that leads to the solution. It makes for great storytelling, but rarely is science like this.

The discovery of Europa's ocean, according to Kivelson, was not a Eureka moment. It came to her and her team after years of chipping away at all the alternative hypotheses. It came after a process of gradually accumulating the right data, as *Galileo* circled Jupiter and flew by its many moons. The data from Europa was critical, but so too were the observations of the other moons, which helped paint a more complete picture of how magnetic fields and moons interact.

To understand the magnetic field measurements made by *Galileo*, it is first worth taking a moment to appreciate the situation at Jupiter and the arrangement of its large moons. Jupiter spins on its axis (makes one full rotation) every ten hours. That is absurd. A "day" on Jupiter is only ten hours long. Jupiter is 318 times as massive as Earth and has a radius nearly 11 times that of Earth, yet it spins more than twice as fast.

Meanwhile, deep within Jupiter the pressure is so high that hydrogen turns into a liquid metallic state. The motion of this core of liquid metallic hydrogen gives rise to Jupiter's large magnetic field. Curiously though, and very fortunately for the study of the moons, Jupiter's magnetic field is not quite aligned with Jupiter's axis of rotation. The axis of the magnetic field is tilted by almost 10 degrees away from the rotation axis. Imagine a bar magnet tilted and then set to spin about its center; as it spins, it carves out two cones, one in the north and one in the south (Figure 5.1).

What this tilt of the field means is that as Jupiter rotates, the magnetic field sweeps past the moons, and the magnetic field is constantly changing because of that slight tilt of the axis relative to the rotation axis. With Jupiter's tilted field, Europa and the other moons see the north pole as slightly closer, and then farther away, as Jupiter rotates over the course of 10 hours. This changing field is akin to the doorway at airport security and it makes for all sorts of interesting interactions with the moons.

The first big discovery made by the magnetometer team was that Io and Ganymede both had strong magnetic signatures. Although embedded within Jupiter's magnetic field, these moons each created local perturbations to the magnetic field that required additional explanation. There had to be something else happening on, or within, these moons.

In Io's case, the extensive volcanism had already been observed, and thus it was relatively clear that material erupting out into space from the volcanoes was disturbing Jupiter's magnetic field. The sulfur and other materials from the eruptions gets ionized, or charged, creating a plasma, which then becomes a conductor disturbing the flow of Jupiter's

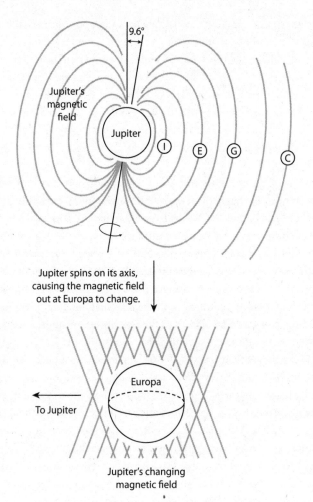

FIGURE 5.1. Jupiter's magnetic field is tilted by 9.6 degrees with respect to its rotation axis. As Jupiter rotates (once every 10 hours), its magnetic field sweeps past its moons and creates an induced magnetic field within Europa. The simplest explanation for this induced field is that Europa has a conductive, global, salty, liquid water ocean a few kilometers to tens of kilometers beneath its icy surface. The bottom diagram shows the magnetic field lines at two different points in time.

magnetic field around Io.[2] Connecting back to airport security, the analogy for Io would be that of walking through the metal detector while spraying gold glitter. Oh, what a scene that would be! The glitter, made of a conductor (gold), would disturb the magnetic field from the security doorway, and the alarm might go off. Glitter-covered security guards would then surely arrest you. Io's incredible volcanic eruptions were spewing material into space that disturbed the magnetic field of Jupiter. In other words, it made sense that the magnetic field around Io would be altered.

But Ganymede presented a more challenging mystery. Why would there be a strong magnetic field around Ganymede? Conventional wisdom held that Ganymede wasn't large enough to have its own field. For a planet or moon to have its own intrinsic magnetic field, it must have at least some form of a molten core. To have a molten core, a world must be large enough to have significant heating at its center from the radiogenic decay of heavy elements and from the formation of the planet itself, that flowing liquid iron can create a magnetic field. Generally speaking, any world smaller than the Earth was considered to be too small to harbor its own magnetic field.

Mars is bigger than Ganymede, and it cooled down long ago, killing off its molten core and any associated magnetic field. Mercury is a bit smaller than Ganymede, and it likely cooled a bit slower given its proximity to the Sun; yet Mercury does not have a strong intrinsic field. Thus Ganymede, it was thought, could not possibly have a molten core and therefore could not have an intrinsic magnetic field.

As a result of this logic, it made no sense for Ganymede to have a magnetic field of its own. Yet a flyby of the *Galileo* spacecraft in 1996 relayed tantalizing indications that, somehow, it did. Measurements revealed that Ganymede had a small magnetic field that was pushing back against Jupiter's incredibly large one.

Although we still have much to understand, Ganymede must have a molten, flowing core of iron to drive the intrinsic magnetic field. Our best guess is that tidal energy dissipation has kept Ganymede pumping along, providing enough heat to maintain the liquid core. Any other viable alternative has proven elusive.

Considering the airport security analogy again, the situation observed at Ganymede would be akin to someone walking through a metal detector while holding a strong magnet in their pocket. Of course, the alarm goes off.

But let's get back to Europa. When *Galileo* made its first measurements of Europa's magnetic field, on December 19, 1996, a peculiar disturbance was observed. It was odd, even when compared to Io and Ganymede's magnetic fields. The field around Europa could not be easily explained by erupting volcanoes or a strong internal magnetic field. These initial measurements hinted at a magnetic field around Europa that rises and falls in direct response to Jupiter's changing magnetic field. In other words, Europa had an induced field, created by Jupiter's magnetic field.

What, then, is within Europa that could create this induced field? In the airport security analogy, you must have a piece of metal or some other conductor in your pocket to set off the alarm. What did Europa have in its pocket that was setting off the proverbial alarm?

Based on the results of the gravity measurements, scientists had good reason to think that Europa has an iron core. Could that be the conductor within Europa? No, it turns out that the core is too small to create an induced field that matches the magnetometer data.

What about a rocky mantle? Again, the gravity data indicated that there should be a thick layer of rocks within Europa, and thus it was reasonable to consider that as a conductive layer that could give rise to the induced field. But rocks, like the silicate rocks that make up our seafloor and mantle, are not very good conductors. Calculations and models for induced fields in a rocky layer within Europa also show a poor match to the data.

It turns out that a global, conducting layer near the surface of Europa matched the data very well. Here again, the gravity measurements and the spectroscopy measurements indicated that the uppermost layer of Europa is water in either liquid or solid form; thus considering water as the conducting layer made sense. When the team plugged conductivity values for water into their models, they found that ice, and even pure liquid water, were not conductive enough to explain the induced magnetic field.

But if salt is added to liquid water, then the conductivity increases dramatically. When the team did the calculations for a salty ocean, they got a good match to the induced field measurements. Beneath Europa's ice shell, a salty liquid water ocean could serve as the conductor through which the time-varying field of Jupiter flows and creates electric currents, which then give rise to the induced field that *Galileo* detected.

But the ocean hypothesis was still very nascent at that point. Images of Europa's surface had sparked the imagination; fractured ice and bizarre iceberg-like structures clearly told the story of something curious happening on and within Europa. By the spring of 1997, the gravity science team had amassed enough data to suggest, still somewhat cautiously, that Europa could harbor a liquid water ocean under its surface (based on the principles discussed in the previous chapter).[3]

Armed with these insights, Kivelson and the magnetometer team started to consider that an ocean could explain the odd signature of the field they had detected. A key additional point that the team began to entertain was that if Europa had a salty ocean, then the conductivity could be sufficiently high so as to create or modify magnetic fields.

One of the key challenges to interpreting the data and reaching any significant conclusion, however, was simply that each flyby was essentially a snapshot in time of the field around Europa. Because of that, with just one flyby, it was not possible to tell what, if anything, was changing and why. More data was needed.

Thankfully, more data was on the way. *Galileo* flew by Europa once again on March 29, 1998, with a closest approach of 1,641 km. The analyses from that flyby were led by Dr. Krishan Khurana, one of Kivelson's proteges and closest colleagues. Khurana had bootstrapped his way through the academic system in India and had earned two PhDs, one in India and one in the United Kingdom. His second PhD was on dynamo theory, which is the theory behind why and how magnetic fields are created. Though he had no real experience studying Jupiter and its moons, he was clearly very bright; Kivelson took a chance and hired him.

Kivelson and Khurana discovered that the data from this second flyby provided a strong indication of an induced magnetic field around Europa. They could rule out a mysterious plasma effect because Europa was well-positioned outside of Jupiter's so-called plasma sheet during this flyby. The data from this flyby matched well with models Khurana and the team had made of electromagnetic induction in a near-surface conducting layer within Europa. From these results, the team concluded that the best explanation for the induced field was a salty liquid-water ocean.

But they were still cautious. Although they felt confident with their data and their mathematical modeling of the results, Kivelson and the team still felt they needed one more piece. The problem was that their data was all from the same basic region and period of the induced field cycle between Jupiter and Europa.

What does that mean? Think of it this way. An induced field, which arises in response to another, constantly changing field (Jupiter for Europa, or the metal detector for the conductor in your pocket), pulses up and down and changes polarity (i.e, North becomes South and South becomes North) in direct response to the changes in that primary field. Their observations showed that as the background field of Jupiter changes during its rotation, Europa's field changes. If Europa's field was an intrinsic field, then the orientation of Jupiter's field would not matter.

Europa's induced field can be imagined as a bar magnet resting in the plane of Europa's equator, which rotates around its center every 11.2 hours—the same time it takes Jupiter's magnetic field to rotate once, relative to Europa (which is slightly longer than 10 hours because Europa moves in its orbit as Jupiter spins). As this imaginary bar magnet rotates, the north and south poles of the magnet circle around the equator. To really see the induced time-varying field of Europa, and the flip from North to South and South to North, what you want to do is make measurements above and below Europa's equator at different times during the rotation of this imaginary bar magnet.

But the flybys to date had all been over the northern region of Europa, and they had happened to occur when the imaginary bar magnet (representing the induced field) was at somewhat similar positions. In other words, they hadn't seen the poles flip yet. That would be the smoking gun for concluding they were measuring an induced field.

Because they hadn't seen the flip yet, Kivelson felt that they could not, with absolute confidence, rule out some sort of odd intrinsic field in

Europa. To measure a flip in the field, if it was there, would require a flyby with the position and timing to detect the induced field coming from the southern part of Jupiter's field.

That opportunity came in January 2000. *Galileo* flew by the southern region of Europa while Jupiter's magnetic field was tilted away from Europa. This configuration, according to the team's prediction of an induced field coming from a salty ocean, should create a magnetic field with the poles flipped, relative to what they had previously measured. If the field was only an intrinsic field, then there should be no flipping of the poles. It was a great prediction, one strong enough to convince the engineering team to adjust the trajectory of *Galileo* to make a flyby at the right time and place to test it.

When the data came in, sure enough, the poles had flipped.

It was not exactly a Eureka moment—years of analyses and refinement had gone into developing the salty ocean hypothesis—but it was pretty gratifying, according to Kivelson. The case for an induced field coming from a salty, subsurface ocean was now nearly closed. It would be hard to refute the data and interpretation Kivelson's team had amassed. Europa has a global, salty ocean of liquid water. For the first time in history, we could look to the night sky and know that oceans exist on and within distant worlds.

This concludes the third piece of the puzzle. By adhering to the physics of airport security and magnetic field interactions, the meticulous magnetometer team had provided the final, and arguably strongest, evidence for a global, salty, subsurface ocean approximately 100 km deep, beneath Europa's icy shell.

The next time you pass through airport security, I encourage you to think about Europa and Jupiter. I never did manage to set off the alarm with my little bottle of salt water. But thinking about Europa's ocean always was, and continues to be, a great distraction from the frustrations of a long security line.

The combination of the evidence from finding a rainbow connection, babysitting a spacecraft, and adhering to airport security all build the case for a liquid water ocean within Europa. Spectroscopy tells of an icy surface, gravity data tells of a thick outer shell of water, and the magnetometer data requires a near-surface conducting layer that is best explained by a global, salty, liquid water ocean. These are the three easy pieces to discovering an alien ocean on Europa. This combined with models for the tidal energy dissipation and Laplace resonance indicate that Europa has been consistently heated by tides and radiogenic decay of heavy elements since it was formed. In other words, Europa's ocean exists today, and it has likely been there for billions of years.

As we will see in the next few chapters, the discovery of oceans beneath the ice shells of other moons in our solar system follows a similar pattern of collecting evidence and piece-by-piece coming to the conclusion that alien oceans exist. In some cases, the pieces of the puzzle are the same (spectroscopy, gravity, and magnetometry). In other cases, additional evidence or methods was critical to their discovery.

CHAPTER 6

LADY WITH A VEIL

Enceladus is a curious moon, and that's saying something because Saturn has a herd of over 50-odd moons of all shapes and sizes. Enceladus is only 504 km in diameter, which is about the distance from Chicago to Cleveland, or from Johannesburg to Durbin. For such a small object, one might reasonably expect that there is not much going on; it should be a cold, inactive moon with a surface pockmarked by craters, indicating billions of years' worth of impacts from stray rocks in space.

And yet it is not a cold, dead, inactive moon. Enceladus is alive with geologic activity, sending jets of salt-rich water hundreds of kilometers out into space. Enceladus is careening around Saturn every 33 hours, and emanating from its south pole is a veil of water vapor and particles. As we explore in this chapter, that veil tells the story of an alien ocean deep within Enceladus.

Long before we knew Enceladus has an ocean, it hinted at a planetary puzzle ripe for exploration. Similar to the story for Europa, ground-based spectroscopy with telescopes revealed that Enceladus was an ice-covered moon. The first clues of an ocean below arrived when images from *Voyager* 2 were received by the Deep Space Network in August 1981.

The images from that flyby showed that part of Enceladus's surface—the northern part—was covered with many impact craters. For planetary scientists, craters are like a crude clock: more craters means more time has passed, since it takes time for craters to accumulate. A surface covered with craters is most likely an old surface. The size and abundance of craters in the northern hemisphere of Enceladus indicated that the ice is perhaps 3.5 billion years old.

But to the south, very few craters were seen. No craters meant that Mother Nature was somehow resurfacing Enceladus. Something was happening on, and perhaps within, Enceladus that caused craters to be erased. Like footprints in the snow covered during a storm, something was generating a fresh surface in the southern part of Enceladus.

These *Voyager* images would be the last images of Enceladus for over two decades. Not until the *Cassini* spacecraft made its first flyby of Enceladus on March 8, 2005, would the mystery be reopened.

The *Cassini* mission was cast in the mold of the *Galileo* mission; after the mission to orbit Jupiter, the next stop was Saturn. Launched in 1997, the spacecraft took seven years to journey out to the ringed planet. Once there, the spacecraft danced around Saturn for over 13 years. With every orbit, it turned its many instruments toward Saturn, its rings, and its moons, each time providing a window into wonders never before seen.

With the *Cassini* spacecraft in orbit, the real detective work on Enceladus began. Like *Galileo*, *Cassini* carried a magnetometer on board. However, unlike Europa, the magnetometer measurements were not able to reveal an induced field indicative of a subsurface ocean within Enceladus. Saturn's magnetic field, although strong, is not tilted, and thus there is not a rapidly changing magnetic field to create an induced field within Enceladus. Mother Nature did not set up something akin to airport security in the Saturnian system.

What the magnetometer did reveal is a distortion in Saturn's magnetic field around the southern region of Enceladus. This distortion was best explained by material coming off of Enceladus and interacting with the magnetic field. The signature was somewhat similar to what *Galileo* saw around Io, where the volcanic plumes caused the magnetic field of Jupiter to be distorted. Enceladus, however, was a world of ice, and the measurements could not be explained by volcanoes like those seen on Io. What then could be disrupting Saturn's magnetic field in the region near the south pole of Enceladus?

On the next flyby of Enceladus, the *Cassini* imaging team, led by Carolyn Porco of the Space Science Institute in Berkeley, California, managed to get the right alignment of the spacecraft, Enceladus, and the Sun, such that sunlight would bounce off whatever might be around the south pole, and the camera could snap a few pictures.

The results were breathtaking. The photos revealed a forest of plumes erupting into space. The questions then arose: What were these plumes made of and where were they coming from?

TASTING AN ALIEN OCEAN

To understand the origin of the plumes, the *Cassini* spacecraft needed to taste the plumes. It needed to fly through the plumes and directly sample the material to see what it was made of. This was the desire of the *Cassini* science team, but it fell on the engineers to make it possible. Sampling the plumes would require sending the spacecraft within 100 km of Enceladus's surface.

To put that distance in perspective, imagine throwing a baseball from Los Angeles to New York City and having it come within six inches of home plate in Yankee stadium. That is what *Cassini* did at Enceladus, made possible by brilliant engineers, like my friend Dr. Nathan Strange, at JPL. Without them, there is no science.

The chemical measurements of Enceladus's plumes were conducted by two different mass spectrometers on the *Cassini* spacecraft: the Ion and Neutral Mass Spectrometer (INMS), which was operated by the Southwest Research Institute in Texas, and the Cosmic Dust Analyzer (CDA), which was operated by the University of Stuttgart in Germany.

Mass spectrometers are incredibly useful and are found in most every laboratory that does any chemistry or geochemistry work. They are like the hammer in a carpenter's tool belt—many tools are desirable, but you'll never leave home without your hammer.

Simply put, mass spectrometers are very good at sorting molecules. If you've ever seen one of those devices that sorts coins, you've seen something that is analogous to a mass spectrometer. Coin sorters send pennies, nickels, dimes, and quarters down a ramp where they roll into different slots and stack up so you can see how many of each coin you have. Mass spectrometers use sophisticated electronics and magnetic fields to sort through molecules sent into the instrument. What comes out is an inventory of the molecules: what type of molecules and how many of each.

This analogy is limited though because there is an added complication. Imagine if you could send a dollar bill (or a five-dollar bill, a ten-dollar bill, etc.) into a coin sorter; but as soon as you put the bill in, the sorter broke it into a combination of coins that equals a dollar and then sorted those coins. The data coming out of the coin sorter would only tell you which coins are present, but it wouldn't tell you that some coins started out as a dollar bill.

In the realm of mass spectrometry, the dollar bill is like a very large molecule that comes into the mass spectrometer and breaks—or cracks—into several smaller pieces, each of which is sorted as a smaller piece. All we get to see at the end is the smaller pieces. Therefore, it is difficult to reliably reconstruct the pieces into the original molecules. It's a bit like having to put Humpty Dumpty's egg shell back together without knowing that Humpty Dumpty was originally an egg.

Furthermore, while mass spectrometers are relatively easy to find and operate on Earth, getting these instruments to work reliably on a spacecraft a billion kilometers away, as it flies through a plume of material at 4–8 kilometers per second is a much bigger challenge. The *Cassini* spacecraft flew by Enceladus at these speeds. Collecting molecules at such high velocity is like taking Humpty Dumpty and dropping him from the Empire State Building. There are so many pieces that reconstructing him becomes nearly impossible.

On top of that, an instrument flown in space is often a much simpler version of the instrument we use on Earth. Because there are so many new challenges in space, some of the complicated measurement capabilities can be cut. For example, on Earth many mass spectrometers can measure very large molecules and can distinguish between two molecules that are similar in mass. The units used are usually atomic mass units (amu). One amu is approximately the mass of one hydrogen atom (roughly equivalent to the mass of a proton or neutron). An atom of carbon, with 6 protons and 6 neutrons, has a mass of 12 amu. Large molecules, like proteins and DNA, can be made of so many atoms that they have masses of thousands to tens of thousands amu. Modern mass spectrometers can measure these huge molecules and can identify slight differences in the mass of large molecules (i.e., they are sensitive to small fractions of an amu).

The INMS, as amazing as it was, was constrained by the resources available on the *Cassini* spacecraft and by the complexity of operating in space. For these reasons, its maximum measurement was 99 amu, and it could only distinguish molecules within a resolution of 1 amu. This meant that the INMS could only detect small molecules, and it could not discriminate well between them. Returning to the Humpty Dumpty analogy: imagine if during your effort to glue Humpty's shell back together, you could only find the smallest pieces of the shell and their edges were hard to distinguish from each other. You'd have a hard time putting the pieces back together and figuring out what Humpty originally looked like.

Thankfully, scientists and engineers at NASA, the Southwest Research Institute, and the University of Stuttgart solved many of the problems associated with such a challenging instrument, and they managed to capture and measure material directly from the plumes of Enceladus.

The initial results from the INMS were tantalizing. The data showed lots of water, and within it carbon dioxide, methane, and small carbon compounds (also known as small organics) such as ethane and propane. On Earth, and within our ocean, carbon dioxide and methane can be found emanating from hydrothermal vents. Small organics can also flow from vents. These results caused many in the planetary science community (including me) to gasp with astonishment. Could this water be coming from an ocean below the ice? Could this ocean be rich with carbon compounds that are useful for—or even made by—life?

As exciting as the initial results were, there was one key astronomical object that stood in the way of an ocean interpretation: comets. Comets are big balls of ice and rock that, like Enceladus, emit jets of water and lots of organic compounds. The tail of a comet that we see in the night sky comes from jets of ice and other materials that sublime off the comet as it approaches the Sun and heats up. Although they have jets of water (from ice), comets do not have liquid water oceans, of course. The organics of comets have been studied for decades, and scientists believe comets formed from carbon that coalesced into ices during the formation of our solar system. These icy objects were then subsequently "cooked" by the ultraviolet light from the Sun and energetic particles zooming through space. In other words, the organics in comets are very interesting—and may even have been important to the origin of life on Earth—but scientists don't think these organics predict an ocean within comets.

The initial INMS measurements of Enceladus's plumes were exciting, but it was possible that Enceladus was similar to a glorified comet. It was possible that Enceladus had a bunch of organic-rich ice that had nothing to do with an ocean or life, and it was jetting this material out into space, similar to comets. I was not involved with any of these measurements, but I was forced to adopt the "glorified comet" hypothesis myself. As excited as I was by the ocean hypothesis, it was still an extraordinary claim, and more evidence was needed.

Thankfully, I did not have to wait long. Another instrument on *Cassini*, the Cosmic Dust Analyzer (CDA) served as the perfect complement to the INMS. Whereas the INMS measured the vapor in Enceladus's plumes, the CDA measured the chemistry of the small ice grains ejected by the plumes. The CDA works by having a fancy bucket that collects material from Enceladus's plumes. The ice and dust careen into the bucket and get smashed apart. Those smashed up parts are then measured by a mass spectrometer. The CDA was not as sensitive as the INMS, but it could measure larger compounds; and since it had the ability to break apart grains of ice, it could measure what was in those grains.

The first big "wow" from the CDA was data showing salts in the ice grains of one of Saturn's rings, the so-called E-ring. The E-ring has long been known to exist very close to the orbit of Enceladus; and thus when *Cassini* found plumes on Enceladus, it was not too surprising to see that the E-ring was sourced by the plumes. In other words, by sampling material from the E-rings and finding salts, the CDA was also sampling the plumes and finding salts in the plumes. Salts are a big "wow" because comets do not have salts (to the best of our knowledge). Salts—like table salt (NaCl), bath salts (MgSO$_4$), and potassium chloride (KCl)—are found in places where liquid water has leached through rocks. Rocks supply salts to water; thus if you observe salts, it is generally a sign that liquid water and rocks have mixed together in some geochemically interesting way. Earth's ocean is salty because of the way ocean water mixes with the rocks of the seafloor, and because of the way in which rivers bring material into the ocean. When the CDA results for salts were combined with the INMS results, the case for a subsurface ocean within Enceladus became a lot stronger. I, for one, started to become a real believer.

Then the story got even more interesting. Nestled within the data from the CDA were some curious peaks in the plots. The team, led at the time by the very meticulous and cautious scientists Sascha Kempf and Frank Postberg, noticed some peaks in the CDA mass spectra, indicating that very small silica particles (SiO$_2$) were coming from the plumes. Within Earth's ocean, tiny silica particles can be associated with a significant amount of heat and interesting geochemistry. Some of the hydrothermal vents mentioned in chapter 1 are great sources of silica. We examine that chemistry in a later chapter, but for now the important conclusion to appreciate is that the silica measured by the CDA implicated not only an ocean within Enceladus, but an ocean with an active seafloor of possibly modest hydrothermal activity.

This was big news from an astrobiology perspective. Not only are hydrothermal vents important for habitability, but they might also be locales for the origin of life itself. If Enceladus has hydrothermal vents, then it might be a good place for microbial life to arise and survive.

The *Cassini* team realized that if the silica particles did, in fact, indicate hydrothermal activity within Enceladus, then a testable prediction could be made about the composition of the plume. Hydrothermal vents, such as those that produce silica, also produce large amounts of molecular hydrogen (H$_2$). The INMS had the capability to detect hydrogen; but in order to distinguish hydrogen coming from hydrothermal vents from stray hydrogen made from the breakdown of larger molecules, the spacecraft would have to make a daring flyby through the plumes with the instrument specially tuned to increase its sensitivity to hydrogen.

So, during the Enceladus flyby in October 2015, the *Cassini* spacecraft dove through the plumes in exactly the right spot and in exactly the right orientation to make it possible for the INMS to search for hydrogen.

The results from that final flyby through the plumes showed that Enceladus had more hydrogen than could be explained by standard molecular accounting; this was not just hydrogen made by breaking apart larger molecules.

Here again the team was very thorough and cautious with their analyses. And yet, they could not explain the data with less extraordinary mechanisms. Perhaps the H_2 was being produced by the radiation processing of water ice from H_2O into H_2 and O_2? Or perhaps methane and other organics were being broken apart, and some of the hydrogen from those molecules was combining into H_2? These explanations, and many more, were insufficient. If water molecules were being destroyed, then there should have been a lot of O_2 along with the H_2. If methane and other organics were the source of H_2, then there would need to be a lot more of those compounds available to produce the abundant hydrogen. In the end, the excess of hydrogen still required an explanation.

The best explanation for the excess of hydrogen was that it was coming from a subsurface ocean and that the ocean has active hydrothermal vents that are releasing the hydrogen into the ocean water. The combination of the salts, silica, methane, and hydrogen found in the plumes of Enceladus all piece together giving shape to a chemically rich ocean with a hydrothermally active seafloor. An extraordinary discovery.

As a final thought on the hydrogen, it is interesting to consider whether or not life within the ocean could feed off that hydrogen. Microbes love hydrogen. Hot springs and hydrothermal vents on Earth that gush out hydrogen are prime places for microbes. They chew up the hydrogen, combine it with compounds like carbon dioxide or sulfate, and harness the energy they need to keep on living and multiplying. Is the amount of hydrogen found in the plumes of Enceladus enough to support life in the ocean below?

The answer seems to be yes. A group of scientists, led by a team in Vienna, showed that the chemistry of Enceladus's ocean—as constrained by the *Cassini* measurements—could support a few kinds of microbes that eat hydrogen and carbon dioxide, creating methane in the process. They grew microbes under similar temperature, pressure, and chemical conditions expected in Enceladus's ocean. The proof of concept exists. The INMS detected methane within Enceladus's plumes: Could it be from the exhalation of busy microbes in the ocean below? It is an extraordinary claim, but one we cannot yet rule out.

A SOUTHERN SEA OR A GLOBAL OCEAN?

The *Cassini* results from Enceladus's plumes provided incredible evidence for a subsurface ocean, but how large is that ocean? Unlike the situation at Europa, where the induced magnetic field implicated a *global* salty ocean, the *Cassini* data of the plumes could only implicate an ocean beneath the south pole. Initially, there was not much usable information to determine how extensive Enceladus's ocean is. Perhaps the plumes were only coming from a transient pocket of liquid water and not from an expansive ocean. The former would be interesting but perhaps not a great environment for life; whereas a large ocean would offer the prospect of a sustained and potentially habitable region within Enceladus.

Gradually, a picture began to emerge. A team of scientists, led by Peter Thomas of Cornell, analyzed images of Enceladus and reported that its ice shell seemed to wobble back and forth more than expected. Wobbling, in and of itself, is not surprising. Moons often have real wobbles, called physical librations, and perceived wobbles, which are an optical illusion and are called optical librations (or geometric librations). Moons often appear to wobble with optical librations if they are tidally locked and have elliptical orbits. Simply put, there ends up being a slight mismatch between the rotation rate of the moon and its orbit.

Our Moon serves as a good example of this kind of wobble. It has a slightly elliptical orbit around the Earth, and it takes the Moon 27.3 days, or almost a month, to make that orbit. When the Moon is closest to the Earth (at its perigee), it is moving faster than when it is farthest away (at its apogee). This is a basic consequence of Kepler's laws.

Meanwhile, our Moon rotates on its axis once every 27.3 days, and that doesn't change throughout the Moon's orbit. Its rotation rate is constant.

As a result, when the Moon is at its apogee (farthest away from the Earth), it is slightly over-rotated because it has rotated on its axis faster than it has traveled on its elliptical orbit. During its perigee (closest approach to the Earth), however, the Moon is under-rotated relative to its motion around the planet.

The result of this slight mismatch is that there is an apparent wobble, or optical libration, of the Moon as it orbits the Earth. Again, the reason it's called an optical, or geometric, libration is because the Moon is not actually librating; it only looks like it is from a viewer's perspective on Earth.

There are, however, actual physical librations—real wobbles that occur as moons orbit their planets. These librations occur when a world, such as our Moon, is not a perfect sphere and gravity tries to realign the mass of the moon with its planet. Our Moon is a little oblate and has a fixed tidal bulge that gets pulled on by the Earth. When a moon has a bit of a distorted shape, the slight mismatch in timing, described for the optical libration, leads to the bulge on the moon being pulled back toward its planet. It's like those old Weebles toys; you can knock them over or spin them around, but they always want to stand upright because they contain a heavy weight in the bottom, which is pulled toward the Earth. Our Moon has a slight wobble because it is always trying to reorient toward the pull of Earth's gravity, but the Moon's spin and orbit never let it rest. It's constantly wobbling.

Optical and physical librations occur for most of the moons in our solar system. When a moon is made of solid rock and ice, we can calculate how much wobbling should occur. If, however, a moon is made of something other than solid materials, then the wobble can be larger. An icy crust floating on a liquid ocean can move around and more easily realign itself with the gravity of the planet. Large physical librations can indicate that the moon's icy shell is separated from its rocky interior by a liquid water ocean.

The measurements made by Thomas and team revealed a large physical libration for Enceladus. The best explanation for that wobble is a global ocean that allows the icy shell above to slosh around. Based on their calculations, the ice shell is perhaps 13 km thick near the south pole, and as much as 26 km thick elsewhere. The global ocean, they estimate, is likely 26–31 km deep.

Meticulous tracking of a tiny wobble made all the difference between a south polar sea and a global ocean on Enceladus.

THE RECORD OF THE RINGS

The data we have on Encealdus points to a habitable, and potentially inhabited, global, chemically rich, subsurface liquid water ocean. Enceladus checks all the boxes and is a great place to search for life.

There is, however, one parameter that confounds me: time. How old is Enceladus and how long has its ocean been around? When it comes to life, time may be a critical ingredient. We do not yet have a complete understanding of Enceladus's history, and it may turn out that it is quite young, at least by solar system standards.

The key clue comes from Saturn's rings. Saturn is famous for its beautiful rings, clearly visible through even a modest telescope. For all the serenity of those rings, they may tell the tale of a tumultuous past. The rings may have been created by a massive collision between two objects tens to a hundred million years ago. Exactly what those objects were and how big of a mess it made, nobody knows. Perhaps it was a dwarf planet careening in from the Kuiper belt, or some other rogue object of rock and ice. Whatever it was, it created the dust and ice that now form Saturn's rings. According to some models, the collision event could not have happened more than a hundred million years ago because the rings are not stable for longer than that; Saturn's gravity is slowly cleaning up the mess, sucking some of the pieces inward and pushing others outward.

Along with the rings, it's possible that the collision disturbed, or possibly even created, some of Saturn's small moons, such as Enceladus. Were that the case, Enceladus could be very young, and all those craters in its ice might be remnants from the aftermath of the collision.

If Enceladus's ocean is relatively young and new, could it still harbor life?

We do not know how long it takes for life to originate. On Earth, life may have originated many times, and it likely got a strong foothold not long after the Earth cooled and started to settle down a bit, some 4 billion years ago. The origin of life could happen quickly, or it could take many millions of years. Either way, it seems a safe bet that more time is better. An ocean that has been around for billions of years may have a better chance of harboring life than one that's only been around for tens of millions of years. On the other hand, tens of millions of years could be plenty of time. *We just don't know.*

The data and models for Enceladus are still relatively new. For the most part, the scientific community is still in the early days of analysis and deliberation. It will likely take years to find every clue in the *Cassini* data that tells us more about Enceladus's history. Enceladus is clearly a great place to search for life beyond Earth—the plumes, the ocean, and the chemistry all beckon for further exploration. As we continue to dissect the data and understand more about Enceladus in the past and in the present, I hope we can simultaneously start building a new spacecraft to head out toward this mysteriously veiled moon and once again sample those tantalizing plumes, searching for signs of life within.

CHAPTER 7

THE QUEEN OF CARBON

In some ways, the landscape of Saturn's largest moon, Titan, might feel oddly familiar and comfortable. During the right time of year, which for Titan is equivalent to about 29.5 Earth years, you might find a stream at your feet, trickling away as it bobbles toward a large and beautiful lake. The shores of that lake might show the subtle ebb and flow of tiny waves crashing on a quiet beach, as a slight breeze pushes across the landscape. If it happens to be a bit overcast that day (a Titan day is equivalent to about 16 Earth days), you might hear the pitter-patter of raindrops on your spacesuit. A stream, a lake, and the soothing sound of raindrops—what more could you ask for?

But these sights and sounds would not be provided by the liquid that we know and love on Earth. It's not water that rains down and fills the streams that flow into the lakes and seas. At a chilly −290 °F, the surface of Titan is much too cold for liquid water.

The sights and sounds of this dynamic world are largely the result of methane, *liquid methane* evaporating and cycling through the atmosphere, clouds, rivers, lakes, and seas. Methane, you may recall, is one of the simplest carbon compounds—one carbon atom bound to four hydrogen atoms. On Earth's surface pure methane is only stable as a gas. On Titan, the temperature and pressure (1.5 bars) are such that the meteorological cycle—the clouds, rain, and weather—is based on methane instead of water. Liquid and solid methane are stable on the surface, and as the temperature and pressure vary through the atmosphere, gas phase methane is stable. Like water on Earth, methane can evaporate from the surface, form raindrops in the sky, and flow or freeze on the surface.

Meanwhile, the ground of Titan is water ice, as are the mountains, valleys, and riverbeds. Liquid methane (and some liquid ethane) is the fluid that carves and shapes Titan's surface. This bizarre combination of water ice and liquid methane makes Titan one of the best places in our solar system to search for life—life that is both familiar and unfamiliar. Beneath Titan's icy shell likely resides a deep ocean of liquid water, which could potentially host life as we know it. However, if life were to exist within the liquid methane lakes on Titan's surface, that life's biochemistry would have to be completely different from anything we've encountered on Earth. We cover the prospects for life on, and within, Titan in the pages to come, but first let's take a look at Titan's curious atmosphere because it is critical to the methane cycle that maintains the lakes of liquid hydrocarbons on the surface.

We do not yet fully understand how and why Titan is the way it is. Titan has a thick atmosphere one and a half times as thick as our atmosphere on Earth. It is composed of approximately 95% nitrogen and nearly 5% methane near the surface. No other moon has an atmosphere this thick. Many *planets* don't even have an atmosphere this thick (Jupiter, Saturn, Uranus, and Neptune have thicker atmospheres, but they are gas and ice giant planets). Titan's atmosphere ranks as the second most dense, after Venus and ahead of Earth, among the rocky planets and moons.

Why does Titan have a thick atmosphere, or any atmosphere, for that matter? If you look around the solar system, it does not make sense. For a world to retain an atmosphere, it must have enough gravity to "hold down" the gases, keeping them from just drifting off into space. The Earth's moon, for example, is far too small to retain a significant atmosphere—so is the planet Mercury and Jupiter's moon Callisto, which both lack atmospheres but are about the same size and mass as Titan. Even Jupiter's moon Ganymede, which is bigger and more massive than Titan, does not have an atmosphere. And yet Titan does.

The reason why Titan has an atmosphere remains a great, unanswered question in planetary science. Some of the leading hypotheses involve gases being released from the deep interior of Titan, either slowly leaking into the atmosphere or rapidly refilling it during large-scale geologic events. Although these ideas make sense, they are based on an extraordinary claim: Titan is a geologically active world, both on its surface and in its deep interior. The presence of methane in Titan's atmosphere may actually be a critical clue.

If Titan had an atmosphere made only of *nitrogen*, then perhaps one could say Titan's atmosphere comes from being in a "Goldilocks zone" for such an atmosphere. The explanation would go like this: nitrogen was abundant when Titan was forming; nitrogen is a very stable gas; and Titan's temperature and pressure retained that atmosphere over a long period. This argument is still a stretch, but there is no chemical reason to think that nitrogen wouldn't last.

Methane, however, is a very fragile molecule, at least when ultraviolet light from the Sun is bearing down on it. Methane is rapidly destroyed by sunlight; and even though Titan resides ten times as far from the Sun as the Earth does, it's still not far enough to stave off this destructive reaction. The methane in Titan's atmosphere should only last a few tens of millions of years, perhaps 100 million years at most. Thus, since we observe methane on Titan today, there must have been a large source of methane pumped into the atmosphere tens of millions of years ago or methane is steadily trickling out from below, gradually replacing the destroyed methane. To resupply its atmosphere and outpace destruction of methane by sunlight, Titan would need to have volcanoes or some other mechanism to release gases from below.

So far, no definitive evidence of "cryovolcanism" (think ice volcanoes!) has been observed, but our data is relatively limited—Titan's thick atmosphere made it impossible to use *Cassini*'s camera to observe the surface, much less any cryovolcanic activity.

An intrepid hitchhiker did, however, pierce the atmosphere and touch the surface, providing some clues to geologic activity along the way. The *Cassini* spacecraft carried with it the *Huygens* probe, built by the European Space Agency. When *Cassini* flew by Titan in 2005, it released the

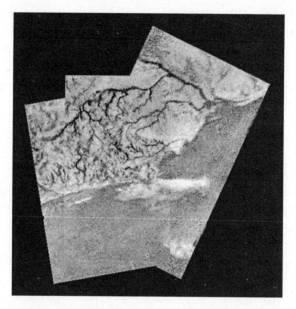

FIGURE 7.1. Titan's surface. As the *Huygens* probe parachuted down through Titan's hazy atmosphere, the surface became visible, revealing an intricate network of river channels. This river network was carved by flowing liquid methane, not liquid water. (Photographs from NASA/JPL/European Space Agency/University of Arizona)

probe and sent it on one of—if not *the*—most remarkable journey a robot (or human) has ever taken. Over many hours, the *Huygens* probe parachuted through Titan's atmosphere, collecting images and data as it descended, eventually landing, quite softly, in a dry riverbed on Titan's surface (Figure 7.1). This riverbed once carried liquid methane, and the rocks that remain are the tumbled stones of water ice that may have been carried downstream by the seasonal flow of methane. The cobblestones of an alien stream are, quite intriguingly, comparable to the familiar rounded ice cubes in a drink left unattended.

As *Huygens* was parachuting down through Titan's atmosphere, it detected the presence of the element argon and at least one of its isotopes. Argon is a noble gas and thus does not form compounds with anything else. For this reason, noble gases are particularly useful indicators of geological and geochemical processes. Argon in its most "natural" form is argon-36, which is made of 18 protons and 18 neutrons. I put the word "natural" in quotes because argon-36 is what stars make during nucleosynthesis, but it is far from the most abundant form of argon we find on Earth. Here we find primarily argon-40, which consists of 18 protons and 22 neutrons.

Argon-40 is not easily made in stars. Instead it forms from the radioactive decay of potassium atoms, specifically potassium-40, which has 19 protons and 21 neutrons (most potassium is potassium-39, which has 19 protons and 20 neutrons). With a half-life of just under 1.25 billion years, one of the protons in potassium-40 converts into a neutron, changing the element to argon-40.

Even though argon-36 is abundant in the atmospheres of our Sun and the giant planets, the rocky planets like Earth formed with so much potassium, including potassium-40, that over time we accumulated argon-40 from the decay of potassium.

So why is the measurement of argon-40 by the *Huygens* probe useful? When potassium-40 decays to argon-40, it goes from an element that likes to be in the solid form (potassium) to an element that likes to be in the gas phase (argon). Argon-40 gas is therefore an indicator of argon being released from outgassing of old rocks and planetary interiors. Rocks within a moon or planet might burp, or leak out, argon-40, as the potassium within the rocks decays.

As a result of this relationship between gaseous argon-40 and solid potassium-40, the detection of argon-40 by the *Huygens* probe was an indication that potassium-40 within Titan's interior had decayed to argon-40. For Argon-40 to be in the atmosphere meant that some form of geological activity had made it possible for argon to reach the atmosphere. Some leak or burp must be coming from below! Is the argon-40 coming from cryovolcanoes? Or perhaps from cracking of the ice shell? We just do not know. But the argon measurement from *Huygens* tells us that Titan has a leak from its interior, and the same leak that's releasing argon-40 could also be sending all that methane into the atmosphere. *Huygens* found a clue that some form of geologic activity is resupplying the atmosphere of Titan.

A TITANIC TILT TOWARD WEIRD LIFE

In addition to being a very scenic world, Titan may also be one of the best places in our solar system to search for life as we know it (water- and carbon-based life) and life as we do not know it ("weird life"). The water- and carbon-based life on Titan would be deep in the subsurface water ocean, which we explore in the next section. The "weird life" would inhabit the liquid methane and ethane lakes that dot the landscape. Those lakes change over time, and the reason they change is because Saturn has seasons.

The seasonal cycle of the hydrocarbon (methane and ethane) lakes on Titan is intriguing both for its fundamental role in shaping the geology of Titan and for its possible role in creating a habitat for weird life. Seasons on Titan allow the winters to be cold enough for huge liquid lakes to form, and the summers are warm enough to evaporate away much of that methane. Seasons may keep the methane meteorological cycle flowing.

On Earth we have seasons because the spin axis of our planet is tilted by 23.4 degrees relative to the plane of our orbit around the Sun. Saturn is also tilted relative to its orbital plane. Its tilt is 26.7 degrees, and the plane of the orbit of its moons share that tilt. For this reason, Titan experiences seasons.

Titan orbits around Saturn every 16 Earth days, and Saturn orbits the Sun once every 29 Earth years. As Saturn and her moons make their orbit around the Sun, the angle of the Sun changes from shining primarily on the northern hemisphere, to shining primarily on the southern hemisphere. This geometry creates the seasons of winter, spring, summer and fall on Titan.

The *Cassini* spacecraft orbited Saturn for 13 Earth years—almost half a Saturn year. During that time, it made 127 close flybys of Titan. When *Cassini* arrived in 2004 and flew by Titan, winter was ending in the north, and summer was ending in the south. The equinox—equal night and day—was in August 2009. As the mission came to an end in 2017, Titan's southern hemisphere was in the depths of winter, while the north was experiencing summer (Figure 7.2).

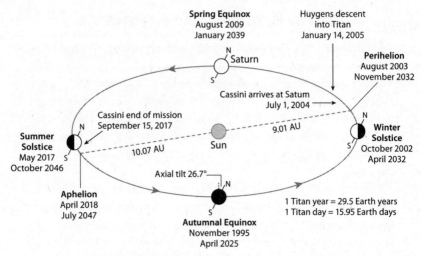

FIGURE 7.2. The seasons of Saturn and its moons, and the timeframe when the *Cassini* spacecraft and the *Huygens* probe made measurements. Because Saturn's rotation axis is tilted by 26.7 degrees relative to the plane of Saturn's orbit around the Sun, Saturn experiences seasons. Most of Saturn's moons, such as Titan and Enceladus, have a similarly tilted axis and also experience seasons, which may cause the lakes, rivers, and rain to migrate back and forth over the year (one Saturn year—one orbit around the Sun—equals 29 Earth years). (Adapted from Hörst, S. M. (2017). Titan's atmosphere and climate. *Journal of Geophysical Research: Planets*, 122(3), 432–482)

When *Cassini* first began flying by Titan, the northern hemisphere had recently come out of the longest days of winter (winter solstice). Using its onboard active radar instrument, *Cassini* was able to see through Titan's hazy, thick atmosphere, down to the surface. The radar revealed a landscape dotted with methane lakes and seas.

However, the radar showed no lakes to the south. There were numerous depressions that might indicate dry lakes, but the landscape in the south had no strong signs of vast liquid methane lakes and rivers. How could that be? Why would a world have such a difference?

It may be that the seasonal cycles on Titan lead to the lakes migrating from pole-to-pole, hemisphere-to-hemisphere, over the course of a Titan year. When it's winter and cold in the north, the methane precipitates out of the atmosphere and pools into depressions on the surface, forming lakes. When it's summer and warm in the north, the methane evaporates and travels through the atmosphere to the cold south, and the methane rains out. The data from the *Cassini* mission were not sufficient to confirm or reject this hypothesis, but it's certainly compelling and elegant. Much like the seasons on Earth that can bring snow, rain, and drought, the seasons on Titan could drive similar wet and dry cycles but with methane instead of water.

Might these lakes host life? When it comes to life, the cold liquid methane within Titan's lakes is probably not a good solvent, or liquid, for life as we know it. Methane molecules are not polar (slightly charged) the way water molecules are. The polarization of water makes it very good at dissolving other polarized molecules and controlling many of the chemical reactions that life as we know it needs. The non-polar liquids of methane and ethane behave much differently. They can dissolve other non-polar compounds, but we do not know what kind of complex chemistry can unfold from reactions in liquid methane. It is possible that liquid methane or ethane could give rise to life, but it would be unlike anything we know, or anything we have so far been able to predict; it would truly be "weird life."

We explore the possibilities of "weird life" in more detail in later chapters, but for now, think about the shores of Titan's lakes and seas and imagine tiny waves lapping onto that shore. I hope that there's even the tiniest speck of life there to enjoy the tides of Titan.

THE OCEAN BELOW

For all of the intrigue and splendor of Titan's surface, and the prospect for "weird life" in the lakes, I am equally interested in what lies deep below.

The clues to a liquid water ocean beneath Titan's surface are similar to those I described for Europa, but they have a slightly different flavor. Recall that the progression for Europa went like this: (1) detect an icy surface, (2) detect a gravity profile that requires water (ice or liquid phase) down to some depth, and (3) detect an induced magnetic field signature that requires a conductive layer (salty ocean) beneath the surface. This sequence applies to Titan, but with some modifications.

First of all, we do not have detailed evidence for the composition of Titan's crust. We think it's water ice, but it has been hard to get a good measurement. Titan's atmosphere confounds our ability to see and sense the composition of the surface. Telescopes on Earth and spectrometers on spacecraft cannot easily see through the atmosphere, and thus good spectra of the surface do not exist. The infrared spectrometer and radar on *Cassini* both provide information at a few specific wavelengths, and those results are consistent with water ice. The *Huygens* probe took a few astonishing images and landed directly on the surface, but it was not able to directly determine the composition of those pebbles and stones seen in the images.

The reality is that nothing other than water ice makes sense for Titan's ice shell; it's not a topic of great debate. Although the evidence is a bit lacking, water ice is the most logical explanation. All the other moons of Saturn have predominantly icy surfaces. Titan's atmosphere may shroud the surface but not our common sense: Titan's shell is made of water ice.

So what about the next piece of the puzzle? What does the gravity data indicate?

The story of Titan's gravity signature is astonishing. Data from the *Cassini* spacecraft revealed changes in the shape of Titan's gravity well over time. Between 2006 and 2011 *Cassini* made six very close flybys of Titan, and with each flyby the Doppler data was collected; slight blue- and red-shifts in the signal sent back from *Cassini* showed the tiny changes in the speed of the spacecraft as it was pulled by Titan's gravity in different ways.

From that data the science team was able to detect more than the moment of inertia, which is an indication of the internal mass distribution (which is what we had for Europa). The team managed to get such good measurements of Titan's gravity field that they could begin to characterize a parameter known as the "k_2 Love number": the ratio between the perturbed gravity "state" and the perturbing gravity "potential."

What the heck does that mean? Simply put, it refers to how easily a world stretches and deforms as the gravity and tides change. The "perturbed state" is the actual stretch and deformation that Titan undergoes as it experiences the tidal stresses from Saturn's gravity. The "perturbing potential" is the total tidal stress from Saturn's gravity that actually exists. That is, the perturbing potential of Saturn's gravity might be trying to stretch and deform Titan more than it can because materials within Titan are resisting that stretch. As a result, Titan's actual perturbed state could be different from the perturbing potential. Think of pulling on a spring: you can pull on different kinds of springs with the same force and they will stretch differently because they might be made of different materials or be coiled differently.

In this case, we're interested in how Titan changes as it orbits Saturn and experiences changes in the gravitational pull from Saturn. For a completely "rigid" solid object, for example, a ball of steel that does not deform at all, the "perturbed state" equals zero and the k_2 Love number equals zero. The object is so rigid that no tidal flexing can occur, and thus it is not "perturbed" at all. Meanwhile, a world made completely of rocks that deform and stretch will have a k_2 Love number of about 0.04. Finally, if you have a completely fluid object, which might easily deform, then its "perturbed state" may actually match, or be larger than, the perturbing potential. In this case the k_2 Love number would be equal to, or greater than, one.

The gravity data from *Cassini* showed that Titan's k_2 Love number is about 0.6, a value too large for a rocky, rigid object that resists deforming from tides. The large value indicates that Titan is deforming, although not completely matching the "perturbed state" of Saturn's "perturbing potential." Titan's perturbed state indicates something highly deformable within Titan. *Titan changes shape as it orbits Saturn, and the best explanation for that change in shape is a fluid ocean beneath its ice shell.* Once again, it's an incredible result from physics.

How thick is the ice and how deep is the ocean? The team that conducted much of this work, led by Luciano Iess and colleagues, concluded that the ice shell is, at most, 100 km thick.[1] As for how deep the ocean is, they cannot say. A number of models that incorporate the gravity data, Titan's density, and some reasonable assumptions about the materials that could make up Titan lead to an ocean that is in the range of 70–100 km thick.[2]

What about the third piece of the puzzle. Are there any magnetic field signatures of a subsurface ocean within Titan? The answer is yes, but it's a different story than the case for Europa.

As the *Huygens* probe descended through Titan's atmosphere, it measured a lot of interesting chemistry and weather-related phenomena. *Huygens* also kept an electronic ear out for any electromagnetic noises coming, for example, from lightning in the clouds.

Sure enough, the *Huygens* data revealed a wealth of information about Titan's atmosphere. But one curious signal stood out. At the extremely low frequency of 36 Hz, the probe showed a persistent signal. On Earth, observations of that frequency are typically a sign of lightning activity. When a lightning bolt strikes, it creates a symphony of both sound energy and electromagnetic energy. Some frequencies—both of sound and electromagnetic energy—travel better than others.

With lightning, the 36-Hz frequency travels very far due to a resonance that is set up between the conductive surface of the Earth (primarily the ocean, lakes, and wet soils) and our ionosphere, which is the conductive part of our atmosphere from about 60 km above Earth to 1,000 km above Earth. This resonance was first predicted for Earth's atmosphere by Winfried Otto Schumann in 1952. Detailed measurements of lightning and confirmation of the signal at 36 Hz were made in the early 1960s. Ever since, this type of resonance has been called the Schumann resonance.

Were the *Huygens* probe measurements of a Schumann resonance indicative of lightning on Titan? At first, that seemed like a logical conclusion. But the images showed no big thunderclouds or storms that could be the source for lightning. The lightning hypothesis fell short.

To have the Schumann resonance, there needs to be a source for the electromagnetic frequencies, such as lightning, and there also needs to be two conducting layers that serve as the boundaries, or waveguide, for the resonance.

Work done by Christian Béghin and colleagues indicated that as Saturn's magnetic field sweeps past Titan, it could provide the source of electromagnetic frequencies into the atmosphere, including the 36 Hz signal.[3] Furthermore, their work showed that Titan's conductive ionosphere could serve as one of the conductive layers for the 36 Hz signal. What about the other conductive layer that is needed?

Titan's hydrocarbon- and ice-covered surface is not electrically conductive; basically it acts as a big insulator. However, the 36 Hz data indicated that approximately 55–80 km beneath Titan's crust,[4] there is likely a conductive layer that serves as the other boundary for the Schumann resonance. *The best explanation for that subsurface conducting layer is a salty liquid water ocean.* Once again, salt helps provide a conducting ocean that can be detected through the interactions of electromagnetic fields.

The team might have been disappointed that they did not find lightning in Titan's atmosphere, but a global subsurface ocean was surely a nice consolation prize!

———

Titan is hands down the queen of carbon in our solar system. Whether in the lakes on the surface or in the liquid water ocean below, Titan likely has all the carbon life might need. The combination of Titan's deep ocean and carbon-rich lakes make it a very compelling place to search for both "weird life" on the surface, and life as we know it (i.e., water-solvent based life) in the subsurface.

One big unanswered question is: How much of the subsurface ocean is brought up to the surface? Recall that the Argon-40 measurements in the atmosphere indicate that at least some of the gases in the atmosphere came from Titan's interior, either recently or in the not-too-distant past. Might some of the same processes that deliver gases from the subsurface also bring ocean water to the surface? Perhaps.

Could "weird life" swim in the seas on the surface of Titan and life as we know it persist in the dark realms of the ocean below? Could microbes in the deep ocean be belching out methane that then serves as the fluid for life on the surface? That could be quite a planetary ecosystem. It's a crazy idea, but it's not out of the question.

For all we know, life could be there right now. One type of creature enjoying a hazy day on the lakeshore, while the other nibbles away at ice and rocks in the ocean below.

Until we get back out there, we'll never know.

CHAPTER 8

OCEANS EVERYWHERE

Europa, Enceladus, and Titan are, in my opinion, the three most compelling ocean worlds when it comes to the prospect of finding life beyond Earth. But many other alien oceans likely exist, and they too, could be habitable or even inhabited. Now that we have established some of the fundamental ways we have come to learn that oceans likely exist beneath icy shells of distant worlds, we can go through, in rapid succession, the evidence for, and astrobiological potential of, other alien oceans in our solar system.

This cast of characters demonstrates how prevalent liquid water may be in the outer solar system. There may be even more ocean worlds out there than I detail below. We do not yet know, for example, whether some of the moons of Uranus, such as Miranda and Titania, harbor oceans.

The total volume of liquid water in all the oceans of the outer solar system could be more than 20 times the volume of liquid water in Earth's ocean. That volume of water represents a tremendous amount of potentially habitable real estate. But as we learned in chapters 2 and 3, liquid water is just one of the keystones needed for life. In numerous cases, the alien oceans described below may pose some serious challenges for providing the elements and energy life needs; they may reside a little too far away from the various Goldilocks zones of chapter 3.

But the universe does not care about our ability to assess the habitability of these worlds. The universe just is. Perhaps life is happily churning away within oceans that don't satisfy our constraints for the origin of life and habitability. All the more reason to get out there and explore.

GANYMEDE

Ganymede is the largest moon in our solar system. It's bigger than Mercury and almost as big as Mars. Ganymede has an ocean, and the evidence is largely derived from the same three pieces of the puzzle that we examined for Europa: spectroscopy, gravity measurements, and magnetic field measurements.

The *Galileo* spacecraft made six close flybys of Ganymede over its lifetime, sending back much of the same kind of data collected for Europa. One interesting exception is that on the fifth flyby, in December 2000, the *Cassini* spacecraft happened to be in the neighborhood. *Cassini* needed to get a slingshot boost from Jupiter on its way out to Saturn, and the gravity assist serendipitously provided the chance to have two spacecraft measuring nearby magnetic fields at the same time in different places. *Galileo* measured Ganymede's magnetic field up close, while *Cassini* measured the background fields of Jupiter and the Sun. This was quite useful, as the magnetic field around Ganymede had confused many scientists.

The story of Ganymede's magnetic field is much more complicated than Europa's. Thanks to the *Galileo* mission we now know that Ganymede has its own, intrinsic magnetic field. The *Galileo* results also showed that there is a time-varying component to Ganymede's magnetic field. Similar to Europa, the time-varying component pointed toward a salty subsurface ocean.

The discovery of Ganymede's intrinsic magnetic field was a very curious finding. Mercury, Mars, and Titan do not have magnetic fields, and they are all comparable in size to Ganymede. Why does Ganymede have its own field? For a world to have a magnetic field it must have a molten region in its core that flows and circulates in a way that creates a magneto-hydrodynamic dynamo (MHD). In computer models of planetary interiors, MHDs look like giant balls of yarn, flowing and moving. Earth has a large MHD flowing as part of its molten iron core, and that motion of liquid iron gives rise to our magnetic field.

Given the magnetic field data from *Galileo*, Ganymede must have a molten, circulating core that drives an MHD, which in turn generates the observed magnetic field. The energy to drive the dynamo within Ganymede is still a great mystery. Theories to date focus largely on tidal energy dissipation, the decay of heavy radiogenic elements, and leftover heat from when Ganymede formed (so-called accretional heat). Together these sources of heat could keep Ganymede's interior hot enough to maintain both a liquid water ocean and an active iron core. Recall that Ganymede is locked into the Laplace resonance, which is that beautiful harmony of orbits where Ganymede orbits Jupiter once for every two orbits Europa makes, and Europa orbits once for every two orbits Io makes. As a result, Ganymede has a forced eccentricity (ellipitical orbit) and must go through periods of increased tidal heating. Whether or not tidal heating from the Laplace resonance is particularly active today, however, is not yet known.

Looking at Ganymede's surface, it's clear that there is a complicated story to be told (Figure 8.1). Roughly one-third of Ganymede's surface is old "dark terrain," while the remaining two-thirds are younger, icy, "bright terrain."[1] The ages of these terrains are estimated from crater counts, with the heavily cratered, dark terrain clocking in at older than 4 billion years on average. The younger, bright terrain clocks in at 2 billion years on average, with some regions as young as 400 million years.[2]

The geology of Ganymede's surface is a testament to nearly every kind of activity—past or present—one could have on an icy moon. Terrains of ice and rock rise and fall 700 meters (2,000 ft), or about twice the height of the Chrysler Building in New York City. Craters of all sizes dot the landscape. Signs of tectonic activity—some familiar, some bizarre—fracture the landscape. Some of the odd, weaving and bending tectonism are called "furrow systems," aptly describing both the geology and the eyebrows of scientists trying to make sense of this confusing surface. Volcanoes of near-freezing water may have paved over patches of the surface with icy magma, although no contemporary activity has been observed in any of the datasets. Perhaps Ganymede was much more active

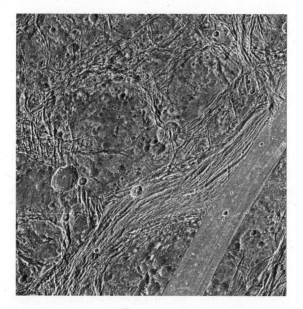

FIGURE 8.1. Ganymede's icy shell is beautiful and geologically confusing. Fractures, bands, and ridges cut through the surface, indicating tectonic activity, likely driven by tides as Ganymede orbits Jupiter. Shown here is Nicholson Regio. The gray band that cuts across the bottom right could be ice pulling apart and new material coming up from below, creating a uniform set of lines in the ice as new ice comes to the surface. The area in this image is 144 km² (90 square miles). (Photograph from NASA/JPL/Brown University)

in the past, or maybe it's on the verge of becoming more active; we just don't know.

The composition of Ganymede's surface tells the story of an old, perhaps salty, ice shell that has collected impacting rocks of all sizes. The dark terrain appears to be a surface made from the debris of meteorites—silicates, clays, and possibly even organics. The fingerprint of Io's volcanism can be found on Ganymede, as with Europa. Sulfur, delivered from Io, coats the surface, where it is subsequently processed into other compounds by the radiation from Jupiter's magnetosphere. Also measured on Ganymede's surface, and in its very faint atmosphere, are oxygen and ozone, two compounds which, if cycled into the ocean, could provide some useful energy for life.

The density of Ganymede, and the interior models made possible from the *Galileo* gravity and magnetometer data, tell a cloudy tale of potential habitability. On the Goldilocks scale for water–rock interactions, Ganymede falls in the range of a large, low-density (1.94 grams per cubic centimeter; g/cm^3) moon that might not offer much in the way of liquid water leaching through rocks to provide the elements and energy life needs.

By mass (i.e., weight), Ganymede is about 60% rock and 40% water in the form of ice or liquid water. The gravity data indicates that the ice and water layer may be approximately 800 km thick. Within that layer, the magnetometer data implicates an ocean of liquid water sandwiched in the ice at a depth of approximately 170 km. The ocean could be tens of kilometers thick to more than 100 km thick. This leads to a volume of liquid water that is perhaps ten times the total volume of liquid water found on Earth.

At the base of Ganymede's ocean, however, is not a rocky seafloor but a form of ice we never see on Earth (unless created in a high-pressure lab). The seafloor of Ganymede is made of ice III, a type of a high-pressure water phase, which is ice made of water molecules that are packed so closely together that the density is greater than that of water. Consequently, ice III sinks. Even deeper in Ganymede's ice layer are likely ice V and ice VI, which again are ices with water molecules pack together in different, denser crystals. On Earth we never see these forms of ice— we only see ice I—because there is no place on Earth cold enough and with high enough pressure to form these curious crystals.

Ganymede's ocean, sandwiched between two layers of ice, could be bad for the chemistry of life. At first glance, it is tempting to conclude that without a rocky seafloor there can be no hydrothermal vents, and without hydrothermal vents, it would be very difficult to sustain life in a dark ocean. I definitely fall on that side of the fence when it comes to the habitability of Ganymede, but my JPL colleagues Christophe Sotin and Steve Vance are quick to point out that there may be pockets, veins, and other dynamics within the ice of Ganymede that cause the deep layers of ice to convect material upwards from the rocky interior.[3] In other words, the icy seafloor of Ganymede's ocean may be cycling with a rocky region deeper within Ganymede. Although this cycling may be less efficient than direct cycling of water with rocks, it could still be enough to keep Ganymede's ocean charged with the chemistry life needs.

Thankfully, within the next decade or so, we will have the chance to see Ganymede up close, once again. The European Space Agency is pioneering a mission to Jupiter and its moons. The *Jupiter Icy Moons Explorer* (JUICE) mission, will fly by many of the large moons before settling down in orbit around Ganymede.

Once in orbit, the mission will be sensitive to all the gravity and magnetic field perturbations, which will reveal much more about the interior of Ganymede. Cameras, spectrometers, and an ice-penetrating radar will show us the surface, the surface composition, and the depth and structure of Ganymede's outermost icy shell. It will help us better understand whether or not Ganymede's surface serves as a window into the ocean below, and whether or not there could be signs of life preserved on that surface.

CALLISTO

In our Goldilocks story of the new habitable zone, Callisto perhaps occupies the edge of the tidally driven habitable zone, somewhat akin to Mars lying out on the edge of the traditional, Sun-driven habitable zone. Callisto is the cold bowl of porridge in the Goldilocks fairy tale. Callisto is about three times farther away from Jupiter than Europa is; and although Callisto's orbit is eccentric (elliptical), it is too far from Jupiter for strong tidal stresses to arise. The tidal tug is at a minimum.

Nevertheless, Callisto does appear to have a subsurface ocean. As with Ganymede, the evidence is again similar to the three easy pieces I described for Europa. Unlike Ganymede, Callisto lacks a significant intrinsic magnetic field, despite Callisto's similar size and mass. Callisto's density is about 1.84 g/cm^3 and its diameter is just over 4820 km; whereas Ganymede's diameter is 5268 km. Like Europa, however, Callisto does have an induced field that necessitates a salty ocean beneath its icy shell.

Once again using craters as a clock to determine surface age, almost all of Callisto's surface falls within the range of 3.9–4.3 billion years old.[4]

FIGURE 8.2. Callisto up close. The Sun shines from the left, and the shadow of a large cliff is cast across the terrain to the right. The area in this image is 33 km (20 miles) across. Each of the bright rimmed craters, large and small, corresponds to an impact. Even the large cliff is a fault in the icy surface caused by one of the largest craters on Callisto, the Valhalla multi-ring impact basin. (Photograph from NASA/JPL/Arizona State University)

That's basically the age of the solar system (4.66 billion years), which means that since Callisto formed and settled down in the Jovian system, its icy surface has not changed much. The youngest regions of the surface are simply more craters that hit Callisto roughly 2 billion years ago. To this end, Callisto is the witness plate for the Jovian system (Figure 8.2). It's the oldest, most cratered icy surface, and it's been collecting those craters, along with everything else, since the earliest days of the solar system. Its surface is a witness to everything that has gone on around Jupiter for the past four billion years.

The images of Callisto's surface from *Voyager* and *Galileo* revealed no evidence for cryovolcanic resurfacing and no obvious signs of tectonics. This is particularly curious given the abundance of such evidence on Ganymede. The surface of Callisto is characterized by craters, plains of dark material, icy regions, and knobs and peaks of various shapes and sizes. The lack of any signs of activity, past or present, on Callisto is consistent with there being very little internal heating. Were Callisto to have more internal heat, then the outer icy shell might be expected to disrupt the surface, refreshing the ice and erasing old craters. Callisto is not part of the Laplace resonance, and thus it has not, to the best of our knowledge, experienced any significant episodic tidal heating similar to Io, Europa, and Ganymede.

The composition of Callisto is broadly similar to that of Ganymede: water ice regions coupled with dark terrains that bear the signature of meteoric infall and impact debris. Silicates, organics, and even some nitrogen have been tentatively identified with spectroscopy.[5] Distinct from Ganymede, however, is the prevalence of ices and frosts of carbon dioxide and sulfur dioxide. Callisto even has a thin atmosphere of carbon dioxide.[6] The sulfur is likely from Io, but the origin of the carbon dioxide is harder to determine. It may be primordial, dating back to the formation and outgassing of Callisto, or it may have accumulated from impacting asteroids and comets. In addition, any organics also delivered by asteroids and comets might gradually be broken down into carbon dioxide by the Sun's ultraviolet photons. Callisto, it should be mentioned, is too far out from Jupiter to experience the intense electron and ion bombardment that Io, Europa, and Ganymede endure.

Moving to the interior of Callisto, the gravity data and models indicate that Callisto is not well-differentiated, but it likely has a mixed rock–metal core that extends out for roughly 600 km. On top of that is a layer of mixed ice and rock, and on top of that is an outer layer of ice as much as 350 km thick. The induced magnetic field of Callisto requires a salty ocean at a depth of 100–300 km within that outer ice layer. The ocean itself may be only tens of kilometers thick.

As with Ganymede, Callisto's ocean is deep, and its seafloor likely resides in a temperature and pressure regime where higher phases of ice, such as ice III, are stable. This could possibly make seafloor interactions with geochemically rich rocky material unlikely. This, again, is not great from the habitability standpoint. Liquid water leeching through silicate rocks is potentially critical for providing the elements and energy that life needs.

Callisto is a truly fascinating world, independent of any prospects for life. When it comes to the search for life, however, the combination of an ocean sandwiched between ice and trapped beneath a thick, old ice shell, means that even if life were to exist there, our chances of ever finding it are exceedingly slim.

For me, and much of the planetary science community, one simple but elegant question motivates our interest in Callisto: Is tidal heating the main difference between how you get Ganymede and how you get Callisto? In other words, if you started pumping Callisto with a bit of a tidal tug, would Callisto start to look more like Ganymede? Might Callisto even start up its own magnetic field?

Experiments like those are obviously beyond the scope of our capability, but hopefully in the future we will collect enough new data on both of these worlds that we can adequately recreate their behavior in computer models. Perhaps in silico (within a computer) we can watch Callisto transform into Ganymede, and vice versa, as we turn the dials on tidal energy.

TRITON

The evidence for something curious happening beneath the ice shell of Neptune's lone large moon, Triton, dates back to the late 1980s. When the *Voyager* 2 spacecraft flew by Neptune during the summer of 1989, it captured images revealing a bizarrely fresh surface with very few impact craters. Parts of Triton's surface drew comparisons to the textured surface of a cantaloupe; nothing like it had ever been seen before on a planetary surface. The only fitting description for those icy surfaces was "cantaloupe terrain."

Coupled with this fresh surface was a sequence of images from *Voyager* that sparked much interest and much debate. The images revealed an active, changing plume of material erupting out of Triton's surface. Like a smokestack billowing out black smoke into a steady breeze, the plumes on Triton cast long, black streaks above the surface and trailed off into the thin nitrogen atmosphere.

These were the first plumes of activity ever witnessed on an icy moon. Sure, *Voyager* had seen the intense volcanism of Io, but that made sense once tidal energy dynamics were understood. An active surface on Triton? Now that was curious. (Keep in mind, this was long before *Cassini* discovered the plumes near the south pole of Enceladus.)

Triton's plumes rise about 8 km above the surface and may be driven by the accumulation of the faint heat from the Sun, in a process typically referred to as the solid-state greenhouse effect.[7] We usually think of the greenhouse effect as something that occurs in atmospheres, such as on Earth and on Venus, where carbon dioxide traps heat, but the greenhouse effect also occurs in solids. On Triton, sunlight may be passing through ices of nitrogen and then hitting dark organic material that is buried within the ice. The dark material absorbs the sunlight and heats up, which in turn causes the ice around it to sublime into gas. Eventually the pressure builds, the ice above breaks away, and an eruption of nitrogen gas and organics occurs.

The plumes of Triton betray some of the more peculiar properties of this moon. The reason energy from the Sun can build up enough to drive the plumes is because Triton has an odd orbit; it orbits Neptune in the opposite direction from Neptune's rotation (called a retrograde orbit). For planetary scientists, this is a huge red flag implying that mischief is afoot. When stars, planets, and moons all form out of a disk of material, it's like a big whirlpool with lots of eddies. Everything in that whirlpool spins and moves in the same direction. If you find an object that goes against the grain, chances are it did not form in place. In other words, Triton's retrograde orbit tells us that it likely formed far out in the Kuiper belt and was captured by Neptune billions of years ago. How exactly that happened, we do not know. But some sort of chance close encounter caused Triton to come under the influence of Neptune's gravity.

The second odd aspect of Triton's orbit is that the plane of its orbit is heavily inclined with respect to Neptune's equator and the plane of Neptune's orbit around the Sun. This large tilt means that parts of the moon experience permanent sunlight, or permanent darkness, for half the Neptune "year," which is nearly 165 Earth years. Right now, Triton's southern hemisphere is bathed in sunlight and the north is locked in dark. The plumes that *Voyager* saw were in the southern hemisphere, and the persistent, slow, steady cooking from the Sun is likely powering those eruptions.

As interesting as the plumes are, they do not actually implicate an ocean beneath Triton's icy crust. Unlike the plumes on Enceladus, which connect to an ocean below, the plumes on Triton are more of a surface and atmospheric process than an indicator of subsurface activity.

The evidence for an ocean is very limited, but even so, it is hard to imagine a scenario in which Triton does not have a liquid water ocean, perhaps with a bit of ammonia mixed in. The young, icy surface of Triton certainly points to vigorous resurfacing that could necessitate a liquid layer beneath the ice shell. Based on the limited number of craters, Paul Schenk, of the Lunar and Planetary Institute, and Kevin Zahnle, of NASA Ames Research Center, calculate Triton's surface to be, on average, younger than 10 million years.[8] By solar system standards, that's the blink of an eye. Triton's thin nitrogen atmosphere might help erode the signature of some craters, but such large-scale resurfacing almost certainly requires convecting ice that cycles from an ocean below. In addition, a number of fractures and smooth, lobed plains on Triton's surface point to tectonic activity and possibly to cryovolcanism that might connect to an ocean.

Beyond the imagery, Triton's composition and history make it hard to escape the ocean hypothesis. Once it was captured by Neptune, Triton would have experienced extensive tides as its initially elliptical orbit spun down to its present-day circular orbit. During that period of orbital evolution, the dissipation of tidal energy within Triton would have generated plenty of heat to create and maintain a global ocean of liquid water.

Today, however, Triton experiences very little tidal dissipation. Its orbit around Neptune takes just under six days, and it's a near-perfect circle. Heating from tides has died down.

Thankfully, at least from an astrobiology standpoint, Triton's composition may have come to the rescue and is serving to sustain a subsurface ocean. With a density of just over 2 g/cm^3, Triton harbors enough rocky material, with enough heavy elements, to generate a slow and steady flux of heat from radiogenic decay. The relatively high density of Triton also bodes well for water–rock interactions that could foster and feed life. Spectroscopically we know that water, methane, and nitrogen ices cover the

Plate 1. A red crustacean enjoying life near the hydrothermal vents at Menez Gwen, 1 km (0.6 miles) below the surface of the Atlantic Ocean. Taken during a dive in the Russian Mir 2 submersible. (Photograph by Kevin Peter Hand)

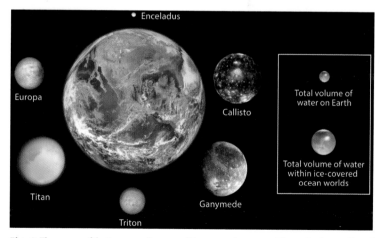

Plate 2. The oceans of the ice-covered moons of the outer solar system. The total volume of liquid water within these six oceans is likely more than 17 times the liquid water in Earth's ocean. (Image credit: NASA/JPL/Kevin Peter Hand)

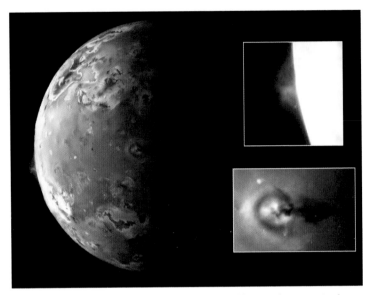

Plate 3. Jupiter's moon Io is the most volcanically active world in our solar system. A volcanic eruption can be seen on the far left, near Io's equator. Upper inset shows the eruption in detail. Bottom inset shows another volcano, with a lava flow extending to the right. (Photographs from NASA/JPL/University of Arizona)

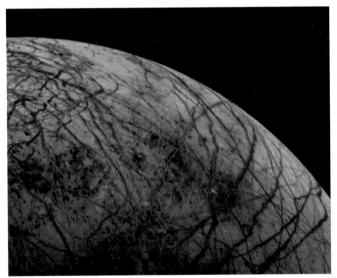

Plate 4. Europa, as seen by *Voyager 2*, July 9, 1979. Although spectroscopy had already revealed that Europa's surface was covered in ice, this was the first time we could see the icy surface up close. (Photograph from NASA/JPL)

Plate 5. Europa, as imaged by the *Galileo* spacecraft in the late 1990s. Fractures cut across the surface, indicating tectonic activity, likely driven by tides. The yellow and red terrain along the fractures may be salt from a subsurface ocean. (Photograph from NASA/JPL–California Institute of Technology/SETI Institute)

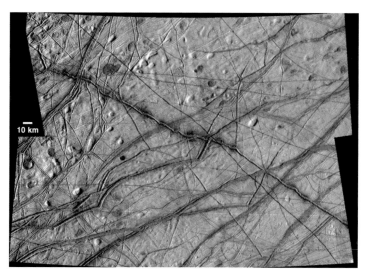

Plate 6. Detail of Europa. These connected fractures can, in part, be explained by the rise and fall of tidal stresses as Europa orbits Jupiter. Regions with dark red material could be the result of eruptions bursting up from an ocean below. (Photograph from NASA/JPL–California Institute of Technology)

Plate 7. Enceladus, from *Voyager 2*, August 1981. The northern hemisphere (right side of image) is heavily cratered, indicating an old, icy surface that has seen many impact events, perhaps over billions of years. The southern hemisphere lacks craters, implying a young surface and some type of geological process that erases old craters and creates new ice. (Photograph from NASA/JPL)

Plate 8. Enceladus, from *Cassini*, October 2008. The heavily cratered northern hemisphere is to the left. The lines covering the surface imply tectonic activity. The canyon of ice (right side) is roughly 1 km (0.6 miles) deep. (Photograph from NASA/JPL/Space Science Institute)

Plate 9. The ocean of Enceladus exhales. Fractures in the ice in the south allow jets of water from a subsurface ocean to be ejected into space. The water is illuminated by the Sun (to the right in this image); from *Cassini*. (Photograph from NASA/JPL/Space Science Institute)

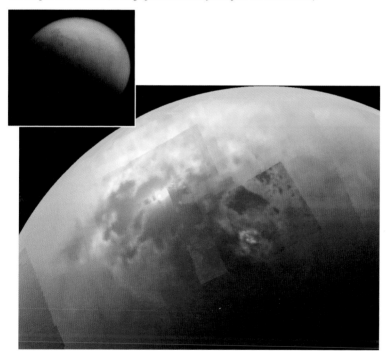

Plate 10. Titan, a beautiful moon of Saturn, has a thick atmosphere of nitrogen, methane, and other organic compounds. The upper image, taken from *Cassini*, reveals dark regions, which are liquid methane seas and lakes. The lower image shows sunlight bouncing off the surface of Kraken Mare (left bright spot), Titan's largest sea. The right bright spot is a cluster of methane clouds that appear orange from the reflected sunlight. (Photograph from NASA/JPL–California Institute of Technology/Space Science Institute/University of Arizona/University of Idaho)

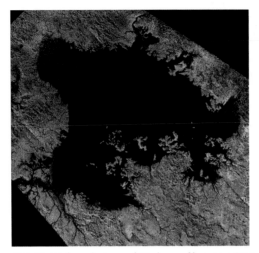

Plate 11. At left, a radar image of Titan's second largest sea, Ligeia Mare (420 km by 350 km; about the size of Lake Superior). This false color image displays the sea and rivers in black and the icy surface in orange and brown. At right, a true color image from the Huygens probe. The "rocks" in the foreground are likely made of water ice, and the surface around them is likely water and methane ice. This may be a dry riverbed, and these rounded rocks (15 cm [6 inches] wide) might be the tumbled remains of a once-flowing river of methane. (Image credit: NASA/JPL–California Institute of Technology/Agenzia Spaziale Italiana/Cornell University/European Space Agency/University of Arizona)

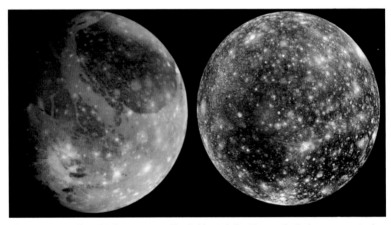

Plate 12. Ganymede and Callisto, as imaged by *Galileo*. At left is Ganymede, the largest moon in the solar system. The brown and gray of the icy surface is a mixture of rocky materials, ice, and possibly salts from a subsurface ocean. The bright spots are exposed fresh ice in the craters. At right is Callisto, which likely harbors an ocean trapped beneath a thick ice shell, billions of years old. As with Ganymede, the dark material on the surface is primarily a mixture of ice and rock. (Photograph from NASA/JPL/German Aerospace Center (DLR)

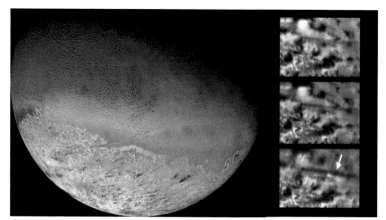

Plate 13. Triton, from *Voyager 2*, 1989. A moon of Neptune, Triton's cold surface ($-390\,°F$) consists of water, nitrogen, and methane ice and frost. The pink regions in the image at left correspond to methane frosts processed by ultraviolet light from the Sun; the green-blue areas (nicknamed "cantaloupe terrain") are nitrogen frosts. The bottom of the image shows dark streaks across the surface, likely active plumes of dust and organic material erupting through the ice. At right is a sequence of images taken 45 minutes apart showing the development of a plume. The top image shows no streak, but the second and third images show a distinct black line emerging; the plume is 8 km (5 miles) high and 150 km (100 miles) long. (Photograph from NASA/JPL/US Geological Survey)

Plate 14. Pluto, from *New Horizons*. Images are color enhanced for contrast. The white and beige regions seen in the image at left are made of water ice, likely mixed with ices of nitrogen, carbon monoxide, and methane. The orange and red regions are likely dominated by carbon- and nitrogen-rich compounds. In the image at right, jagged mountains of water ice can be seen rising several kilometers (the area in image is 80 km [50 miles] wide). These mountains, in the southern part of Pluto, merge with smooth plains of fresh water, nitrogen, and carbon monoxide ices. (Photograph from NASA/Johns Hopkins University-Applied Physics Lab/Southwest Research Institute)

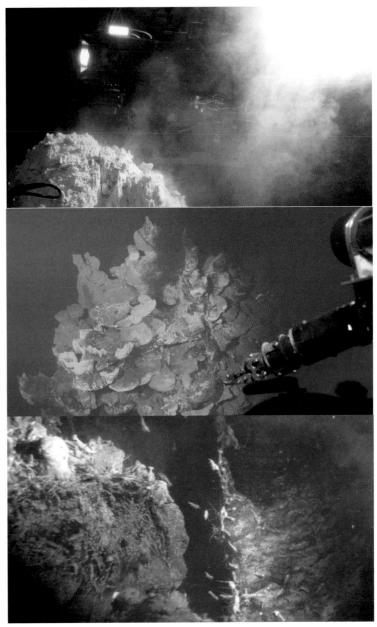

Plate 15. Hydrothermal vents in the depths of Earth's ocean. The top image shows the Lost City hydrothermal vent site, 1 km down in the Atlantic Ocean, which hosts white chimneys of carbonate rock; a submersible hovers as the team collects a rock sample. The middle and bottom images show the Snake Pit hydrothermal vents, 3.6 km beneath the Atlantic Ocean's surface, which feeds a vibrant ecosystem, including microbes, shrimp, and fish. Bowl-like structures host thousands of baby shrimp (in red); adult shrimp are visible in white. (Photographs by Kevin Peter Hand)

surface, and if those compounds are also cycling with rocks in the interior, Triton could be home to a rich stew of an ocean that is ripe for life.

PLUTO

Planet or not, Pluto is pretty amazing.[9] On July 14, 2015, the *New Horizons* spacecraft flew by Pluto after a nine-and-a-half-year journey. Moving faster than any other spacecraft in history, *New Horizons* came within 12,500 km of Pluto's surface. With that singular flyby, the instruments on board collected 50 billion bits of data, each of which took 4.5 hours to get from the spacecraft back to Earth, traveling at the speed of light. Pluto is, on average, 40 times as far away from the Sun as the Earth is. The Sun looks like the head of a pin from Pluto, and the faint sunlight that bathes Pluto gives it an eerie dusk feel, even at high noon. The cameras on *New Horizons* had to be specially tuned and calibrated to make sure they collected enough light to create sharp images.

Pluto is a world of ice, but not just one kind of ice. Water ice, methane ice, carbon monoxide ice, and nitrogen ice dominate the surface geology of Pluto, creating a landscape full of mountains, flowing plains of ice, and possibly even active volcanism. Water ice likely forms much of the strong "bedrock" of Pluto, while the other types of ice flow and erupt on top of it. Meanwhile, as Pluto slowly trots along its 248-year orbit around the Sun, it starts to warm, and some of the nitrogen, methane, and carbon monoxide ices sublime away and form Pluto's very thin, tenuous atmosphere.

On the surface, tectonic features and scarps several kilometers high cut across the terrain in some regions, and mountains built of water ice rise 2–3 km in other regions. The large heart-shaped feature that dominates Pluto's surface is called Tombaugh Regio, named for the discoverer of Pluto, Clyde Tombaugh. Within Tombaugh Regio is a vast region of icy plains called Sputnik Planitia. The plains—rich with nitrogen, carbon monoxide, and methane ices—are covered with puzzle-shaped polygons separated by trenches.

Although many craters can be found on Pluto, one of the most striking first impressions from the *New Horizons* images was the lack of craters. Instead of a long dead, icy relic stowed away in the attic of the solar system, Pluto's relatively "fresh" surface indicated a world of ongoing activity. Indeed, the region of Sputnik Planitia may be the result of a giant impact, whose crater was largely erased by resurfacing and flowing ices.

Many of the most recent models for Pluto indicate that an ocean of liquid water may lie beneath the ice of Sputnik Planitia.[10] The evidence for this ocean has little to do with the three easy pieces described for Europa. Rather the evidence comes from the observation that Sputnik Planitia is a low-lying plain that is, on average, 3 km below the average surface elevation of Pluto. In other words, Sputnik Planitia is not just a plain; it is a basin. In addition, Sputnik Planitia is oriented directly opposite Pluto's largest moon, Charon.

This is suspicious from a planetary standpoint. There are no strong tides between Pluto and Charon, but they orbit each other and are synchronized such that they always show the same side to each other. When this happens, the heaviest (that is, the most massive) parts of a planet or moon end up orienting along the line that connects the two worlds. In other words, if there are any heavier, more massive regions of Pluto, they would eventually be reoriented along the line that connects Pluto and Charon. In general, there's an equal chance for the massive region to end up facing toward Charon or away from Charon.

The curious case of Sputnik Planitia is that as a basin, one would expect it to be less massive than other regions. To explain the alignment of Sputnik Planitia there must be something beneath the ice that gives it a significantly larger mass than other regions. When Francis Nimmo, of the University of California Santa Cruz, and colleagues ran the numbers, they could not solve the mass anomaly problem using other candidate ices or rock. (Note that Pluto's density of roughly 1.86 g/cm^3 indicates very little rock in its interior.) The best explanation is that a liquid water ocean, possibly mixed with ammonia, resides beneath several tens of kilometers of ice in Sputnik Planitia. Liquid water is denser than water ice and fits the observations well.

To be clear, the evidence for an ocean within Pluto is considerably less robust than the evidence for oceans within worlds like Europa and Enceladus, but the data for Pluto point strongly to an ocean.

How could an ocean exist within Pluto, given that there is little solar energy and no tidal energy? Decay of radioactive elements occurs within Pluto and could supply much of the needed heat. But it's a small world and there isn't enough rock to supply the radiogenic heating needed to sustain an ocean.

The key is perhaps in the nature of the ices. Surface temperatures on Pluto range from 40 to 50 K (−390 to −370 °F), depending on how close Pluto is to the Sun in its orbit. At these low temperatures, the ices mentioned above not only form, but they also combine with water to form clathrates. The word "enclathrate" means to put something in a cage, and clathrates are basically tiny, cage-like structures made of water molecules. These cages trap molecules like methane. Importantly, clathrates serve as a much better insulating material than water ice. Clathrates are a great blanket to keep a world warm. As a result, the small amount of radiogenic heat that Pluto generates from its limited supply of rocks can be efficiently retained within Pluto, instead of being lost to space. It is this heat that, in theory, maintains Pluto's liquid water ocean.

From a habitability standpoint, Pluto is very intriguing. Although it may not have much hydrothermal activity or cycling of water with rocks, the inventory of carbon and nitrogen compounds automatically make Pluto of prime interest, in my astrobiological opinion. Were Pluto found to harbor life, I would almost feel comfortable claiming that life is everywhere. Worlds like Pluto should be ubiquitous in the universe; thus if life arises on and within Pluto, then nearly every star should have one or more habitable, or inhabited, Pluto-like worlds full of life.

INTERSTELLAR OCEANS?

Finally, it is important to mention the very real possibility that alien oceans could be drifting through the cosmos, traveling from star to star, solar system to solar system. Professor David Stevenson of Caltech first proposed the idea of habitable "rogue planets" back in 1999,[11] and the idea could be equally well-applied to ice-covered moons with oceans.

Simply put, in the giant gravitational dance done by planets and moons in any solar system, there is a chance that any planet—or moon—could be ejected. As the orbits of moons and planets are working to stabilize in the first billion years or so, a small gravitational tug in the wrong direction could set a world on a chaotic path that eventually leads to sling-shotting around a giant planet and increasing its speed so much that it escapes the solar system.

Once ejected, such a world would be a lonely traveler in the depths of space for millions to billions of years. During that time, it would cool down and potentially freeze all the way through. But a planet or moon with a large rocky interior—and thus an abundance of radiogenic elements—might be able to generate enough heat to sustain an ocean as the rogue world travels the depths of space. Eventually, such a world might come under the influence of a star, and its trajectory might send it slowly inward to join a new family of planets, in a new solar system. At that point, heat from the star, or tidal energy from orbiting a giant planet, could re-ignite the rogue world and restore its biosphere to full bloom.

Like giant interstellar spaceships, such rogue worlds could potentially transport biospheres across the galaxy. Instead of bipedal teams of engineers working hard to keep a massive spacecraft on track, these worlds could haphazardly host microbes, fish, squid, and octopi as they unknowingly travel from one star to another. Perhaps one is even on its way toward us now.

PART III

THE JOURNEY FROM HABITABLE TO INHABITED

CHAPTER 9

BECOMING INHABITED

I have never liked the word "inhabited," yet it is central to the work I do.

We are searching for habitable words in part with the hope that they might be inhabited. If a world like Europa is inhabited, then we stand a chance of finding life there: an inhabited world has life.

But the prefix "in" confuses me. An object that is indestructible cannot be destroyed, a person who is incapable is not capable, and an inanimate object is not animated. Yet an inhabited world is not only habitable, it has life. Alas, it is the term we use.

Grammatical nitpicks aside, a key distinction is that the discovery of a habitable world does not automatically mean that the world is inhabited. The origin of life itself is a bottleneck that we do not fully understand. Is the origin of life easy or hard? Does life arise under many different conditions, or is there only one set of extremely specific conditions under which life will arise? If the origin of life is hard, it does not matter how habitable a world may be; it could be devoid of life because the conditions for the origin of life were never satisfied. In other words, a world could be plenty habitable, but uninhabited.

The ocean worlds of the outer solar system present several possible locations that could be habitable. But that's only part of the question. We want to know if they're *habitable* as a first step to knowing whether they're *inhabited*. We need to know not only where to look, but also what we're looking for and how to recognize it. This raises deep questions about what life is and how it starts. The exploration of ocean worlds will allow us to test different hypotheses about how life itself began. Such exploration helps us better understand where we came from.

As just one example, life on Earth may have begun in warm pools of water on the shores of an ancient ocean, or it may have begun in the bowels of a hydrothermal vent, deep within Earth's ocean. Perhaps it arose in both places, or perhaps in some different locale.

Whatever the case, ice-covered alien oceans offer an important point for comparison. If we do find life in these distant oceans, it would likely have originated within hydrothermal vents, or possibly via some yet to be understood mechanism within the ice shell. Dry land and continents do not exist on these moons, and thus the warm pond scenario for life's origin is not an option. If the origin of life *requires* a warm pond bathed by sunlight, then we will not find life within any of the alien oceans of the outer solar system. If we do find life within these oceans, then we might well conclude that the origin of life within hydrothermal vents is viable, and that may have been how life on Earth arose. The discovery of life—or the lack of any evidence for life—within alien oceans will help inform our understanding of how life on Earth, and ultimately how we humans, came to be.

Throughout the previous chapters we examined some of the Goldilocks conditions for habitability and the science behind why, and how, these potentially habitable worlds exist. Europa, Enceladus, Titan, and possibly several other moons in the outer solar system have liquid water and the essential elements and energy needed to build and power life.

However, we have not yet examined the origin of life itself. Numerous vast global oceans may exist in our solar system; but if the origin of life is hard, then these worlds could be devoid of life. In this chapter and the next we explore some of the conditions required for the origin of life as we know it, and some of the locales on Earth that may provide clues as to whether or not ocean worlds could have given rise to life.

ORIGINS AND HABITABILITY

The origin of life is distinctly different from habitability. A world that is habitable could have the conditions needed to support life but not necessarily the conditions needed for life to originate. Conversely, a world that has supported the origin of life is, by definition, habitable, at least for some period of time (planets change over time, and a habitable world could become uninhabitable).

In general, I think that the conditions for the origin of life are likely harder to come by than the conditions for habitability. The keystones for habitability that we covered in chapter 3 include access to liquid water, access to the elements that serve as the building blocks for life, and access to some form of energy that can serve to power life. The origin of life likely needs all three of these keystones, plus at least two additional factors: a catalytic surface—such as a mineral, where some of life's earliest reactions can take place—and time.

I'll admit that the last parameter, time, is a bit of a fudge factor. How long does it take to go from simple molecules to an organized bag of molecules we can justifiably call life? We have no idea. If the conditions are right, perhaps it only takes minutes. Or perhaps there are so many chemical permutations that need to be explored that it takes a few hundred million years.

The only useful constraint we have is the record of life on Earth. Our planet is 4.66 billion years old, and evidence for ancient microbial life can be found in rocks that are 3.4 to nearly 4 billion years old. There is significant debate about the strength of the evidence for life in the oldest of those rocks, but we can nevertheless conclude that life arose on Earth within the first half billion to billion years.

This is likely a conservative estimate of when life began because for us to locate evidence of 4-billion-year-old life within rocks means that life must have already been robust and widespread throughout Earth; it wasn't just beginning 4 billion years ago. In other words, it was far past the point of origin, and it was now on to spreading and propagating around the planet.

It is also possible that life arose on Earth many times, in many different ways, and was wiped out by large, impacting asteroids and comets—a process that Chris Chyba likes to call the "impact frustration" of life. Perhaps the limiting factor for the origin of life on Earth was not the chemistry itself, but rather the time needed for our planet to settle down.

Given the uncertainty about the time needed for life to arise, I hedge on the safe side and argue that more time is better. The longer the keystone conditions are satisfied, the more likely life will have a chance to get a toehold. This is in part why I slightly favor Europa above Enceladus when it comes to the search for life. We still do not know how long Enceladus's ocean has been around, but we think Europa's ocean has been in existence for much of the history of the solar system. Europa's ocean may provide a longer term, more stable environment for life.

The other keystone for the origin of life is the need for a catalytic surface. A catalyst is a substance that enhances chemical reactions without actually being part of the reaction itself. Think of a catalytic converter in an automobile. The purpose of a catalytic converter is to reduce the polluting emissions, such as carbon monoxide, oxides of nitrogen (NO_x), and small hydrocarbons that result from incomplete combustion in the engine. The catalytic converter is basically a ceramic honeycomb coated with metals such as platinum and palladium. As the engine exhaust flows through the honeycomb, the gases interact with the metal-plated walls, and the bonds weaken just enough to cause reactions to occur that convert the gases into water, carbon dioxide, nitrogen (N_2), and oxygen (O_2). While carbon dioxide is still a bad thing to emit, the end result of a catalytic converter is much better than what went into the converter.

When it comes to the origin of life, the conventional wisdom is that there was likely some sort of catalytic converter in nature that enhanced the reaction needed for life to start. This could have been minerals in the deep ocean that reacted with carbon dioxide to form larger carbon compounds, or it could have been minerals in a tide pool on the shores of an ancient ocean, where evaporation and sunlight might have caused small compounds to bind together forming larger molecules that eventually lead to life. Catalysis provides a template for what eventually become the molecules, metabolism, and structure of life.

Beyond catalysis, what are the steps needed for the origin of life?

Not surprisingly, there is significant debate in the scientific community as to what the origin of life requires, and where and how life on Earth originated. There is general agreement, however, that at least a few key attributes must come into the process early on. These attributes include compartmentalization, a mechanism for information storage and replication, and some form of early metabolism, i.e., a way to "power" life. Let's look at each of these one at a time.

Life, at the most basic level, needs a compartment. Life separates itself from non-living things by building a compartment. On Earth we call these compartments cells, and they are central to life. From the smallest microbe to the largest animal, cells are the fundamental unit of life. Cells separate themselves from their environment by building a membrane, typically made of lipid (fat) molecules.

Compartmentalization must have happened in the early stages of the origin of life, as it is a fundamental quality of life that it is distinct from its surroundings. If we could magically travel back in time and watch the origin of life as we know it, the critical event that might have happened is a single-celled microbe clutching onto a rock (never mind that we would need a microscope to see it). Biologist and pillar in the field of the origin of life, Professor David Deamer, of the University of California Santa Cruz, has shown that some of those early compartments might have been similar to soap bubbles. Soap-like foam would have been readily available on early Earth; from gurgling hot springs to foamy patches on the surface of the ocean, bubbles and films that could have been a useful container for life would have been everywhere. Compartmentalization, however it occurred, must have happened early in the path leading toward the origin of life. Ultimately, if we could travel back in time, we would know early life on Earth by seeing it as separate from a non-living thing, such as a rock.

Information storage and replication are two more key attributes, and they are closely linked. Part of what makes life "life" is that whatever form it takes, it reproduces itself. Mammals mate and give birth; plants drop seeds and beget new plants; microbes make new microbes; and at the base of it all, cells multiply. In order for this to happen, there must be some form of "architectural blueprint" that can be copied to reproduce the cells. This architect, or blueprint, is the information molecule.

For all life on Earth, this blueprint molecule is DNA (deoxyribonucleic acid). Replication via some form of molecule that stores information is likely a universal attribute for life. In our case DNA is a very specialized molecule, and there may well have been several simpler information molecules that kickstarted the origins of life. In other words, DNA might have had some predecessors.

In its earliest form, the information molecule for life as we know it may have been a form of ribonucleic acid (RNA), or even something as generic as a polynucleic acid (PNA). Whatever it was, it may have served both the roles of architect and foreman. What I mean is that DNA is like the architect with the blueprints; all of the information for what to build is there, but the architect never picks up a hammer. DNA serves a similar role in biology. It is essential—as is the architect for a building—but to actually build the structures of life, DNA only doles out information, such as which proteins to build, instead of doing any of the physical work. Meanwhile, RNA functions like the foreman on a job site. The foreman talks with the architect and reads the blueprints and then shows everyone else what and how to build the structures needed.

It's this quality of RNA—being good with both information and action—that led a number of scientists, most notably Leslie Orgel and Gerald Joyce at the Salk Institute, to hypothesize an RNA-world as a key stage in the origin of life. In this model, RNA serves as both the architect and foreman for early life on Earth. It makes sense that both functions would have been centralized in one molecule. Later, as life got going and evolution could accommodate more complexity, the two jobs were split among the two molecules, DNA and RNA.

How one arrives at that early form of RNA is another big problem in the origin of life, but it was perhaps preceded by simpler strands of polynucleic acids (no ribose sugars helping to form the backbone). Many researchers seem to agree that once you get to the RNA-world, it's off to the races with evolution. But getting to that point is still tricky.

Making some form of RNA, or pre-RNA, requires the basic building blocks of life—the nucleic acids, sugars, and amino acids from which the larger molecules are built. These are the "bricks and mortar" of life, if you will. Simply making these compounds is not that difficult. But linking them together in an ordered fashion is hard.

The first modern-era experiments that attempted to address the origin of life were the now classic "primordial soup" experiments of Stanley Miller and Harold Urey. Miller, Urey, and a host of talented chemists showed that making the basic building blocks of life—compounds like amino acids and nucleobases—is not that hard. Mixing together some water, ammonia, and methane can, with the addition of an electric spark or other form of energy, generate a subset of these compounds.

These types of experiments have been replicated under a variety of conditions relevant to early Earth and astrophysics; and although there are some caveats, the general lesson is that the building block compounds of life are not that hard to make. Importantly, many of these compounds (e.g., amino acids and sugars) are found in meteorites and comets, further corroborating the idea that the building blocks are easy to come by.

The difficulty that I have with the primordial soup theory is that there's no real "motivation" to put it all together. In nature molecules don't just link together in an ordered fashion unless there is some sort of geochemical driver. This problem is what my colleague Chris Chyba and I refer to as the top-down versus bottom-up challenge. Taking a top-down approach to life's origins, we can deconstruct the machinery until we get to an RNA-world scenario; but once you deconstruct things beyond the RNA-world, you no longer have sufficient molecular machinery to grow and reproduce. From the top-down perspective, it's hard to make the chemical leap from no life to an RNA-world.

Conversely, the bottom-up approach shows that it's easy to make the building blocks, but putting them together into larger, more functional molecules has proven very challenging in the lab. No one has yet synthesized nucleic acids and sugars and subsequently synthesized those into any viable pre-RNA molecules. There is a large gap between what we know about the origin of life from the bottom-up and top-down approaches. Part of the answer to what lies in the gap is one of the three keystones mentioned above: access to some form of energy that can serve to power life. Or in one word: metabolism.

THE CIRCUIT OF LIFE

In the natural environment there are all sorts of chemical reactions waiting to happen: rust forming on your car, gas waiting to be burned, wine resisting the urge to turn to vinegar, to name a few. Life feasts on all sorts of reactions, and that feast—which is sometimes a famine—is more formally referred to as the metabolism of an organism. At the heart of the origin of life must have been an early form of metabolism—a sequence of chemical reactions that helped life harness energy from its environment. When we think about a planet's or a moon's ability to become inhabited, we can characterize that world in terms of the metabolisms it can support and the energy that can be extracted from the chemistry of the environment. That stored energy is a measure of the chemical disequilibrium in the environment.

Ultimately, life alleviates chemical disequilibrium in the environment. A good example of chemical disequilibrium is a battery. Batteries are a neat package of stored energy made possible by the chemical disequilibrium between the electrolyte (typically an acid) and metal electrodes within the battery. As long as you don't connect the ends of the battery, the energy stays stored for later use.

Our environment is full of chemical disequilibrium. Wherever you look there is stored chemical energy, like little batteries in nature, and there are reactions just waiting to happen. Sometimes those reactions happen dramatically, such as when you light a gas stove and methane burns with oxygen in the air. Other times the reactions take longer, but still have a big impact, such as when your car rusts.

From microbes to humans, biology harnesses those chemical reactions "trapped" in the environment. The metabolisms that drive life accelerate reactions in the environment, *releasing energy faster than would have occurred without life*. Life, in many ways, is a layer on top of the processes of geology and chemistry. At the planetary scale, I like to imagine planets as big geochemical batteries with lots of stored chemical energy that life can use in a variety of ways.

Continuing with the analogy to batteries, the metabolisms of life can, in many ways, be thought of as an electrical circuit connecting the terminals of a battery. The rocks and chemical compounds in the environment are like unused batteries; they want to react but don't have the right "circuit" connected. If a battery is not connected to a circuit, the battery can hold onto its stored energy for many years. If, however, you take that battery and put it into a toy car, and turn the car on, the batteries will run down much faster. By flipping the switch and completing the circuit, you have made it possible for the electrons in the battery to flow through the circuit—causing the car motor to run—and discharging the battery. The energy stored in the batteries is released much faster than if the batteries were just sitting in your cupboard. In exchange for the faster release of energy, you have run the car and gotten something useful out of the energy stored in the batteries.

Biology completes the circuit, so to speak, releasing energy that is stored in the rocks, air, and water of its environment. Like the batteries, the stored energy in the environment would eventually be released without the help of biology, but biology serves to accelerate the rate of release and progression toward equilibrium.

The circuit in biology is the metabolic pathway that life uses. For humans, or for any animal on Earth, our metabolism involves bringing together molecular oxygen (O_2) with some form of carbon compound (e.g., the carbohydrates that we eat). It's basically the same reaction that occurs when burning firewood in a campfire (firewood is mostly cellulose, which is just a large carbon compound that is hard for our bodies to break down . . . which, by the way, is why we don't eat trees). Within our bodies, the "campfire" burning with oxygen and organic compounds is a more carefully controlled "burn."

The technical nomenclature for the compounds used in metabolism and the circuit of life are oxidant and reductant. The oxidant is comparable to the positive terminal of the battery, and the reductant is like the negative terminal. Oxidants want to gain electrons, and reductants have electrons to share, similar to the terminals on a battery. The oxidant for humans is oxygen. The reductants are the carbohydrates, sugars, and other organic compounds we eat.

For all animals on Earth, the energy circuit of life feeds off a "battery" made from oxygen and the foods we digest. The metabolism of animals—including humans—is like a 12-volt battery in your car: there's a lot of energy available in the reaction between oxygen and hydrocarbons.

For microbes, planet Earth offers many different geochemical "batteries" to complete the circuit of life. Microbes can use an incredible range of metabolic pathways, bringing together any number of compounds in their environment that serve as oxidants and reductants. A few of their favorite oxidants include sulfate, carbon dioxide, oxygen, peroxide, nitrate, and iron and manganese oxide minerals. Reductants that microbes love to eat include hydrogen, methane, hydrogen sulfide, various organics, and a host of different minerals. Microbes can harness energy across a wide variety of geochemical conditions, combining metals and salts and heavy acids and bases—you name it, and chances are there's a microbe that knows how to eat it. Considering the battery analogy again, microbes have figured out how to make "batteries" large and small out of nearly every geochemical environment we've explored on Earth. Some of the chemical pathways that microbes use are analogous to the tiniest watch batteries—low power with not much energy released. But that's all the microbe needs to sit there and survive. It's a tiny circuit churning away, releasing energy from the environment to do the business of life.

MIND THE GAP

Returning to the gap between the top-down and bottom-up approaches to the origin of life, while compartmentalization and replication are important, they are aspects of *what* life is and does, but they do not address the *why* of life. The *why* of life is metabolism. By completing the circuit of life, biology harnesses energy from its environment. Technically speaking, this means that biology actually helps the universe cool faster; it increases the entropy of the universe. This is why the universe needs life.

To that end, it may be that metabolism is at the heart of why life gets started in the first place. Metabolism may be the key to closing the gap between the top-down and bottom-up approaches to the origin of life. As my colleague Everett Shock of Arizona State University likes to say, the link between geochemistry and metabolism is like a free lunch that you are paid to eat. The chemical disequilibrium trapped in the environment wants to be released—it's free—and if you can figure out how to release it, you can use it to your advantage to build the structures of life, i.e., you get paid to eat it.

For example, building large molecules requires energy. Synthesizing the molecules of life, such as a pre-RNA molecule or the molecules of cell membranes, requires some form of chemical disequilibrium in the environment. As life was getting started on Earth, the geochemical driver for those early reactions was likely some form of a mineral surface that served as a catalyst for the synthesis of larger molecules. When small molecules attach to metals or minerals, it becomes energetically easier for them to combine with other molecules. This catalytic property of minerals is used in all sorts of ways in the chemical industry, and there are several ways it may have played a role in the early proto-metabolism of life. Although there are many plausible hypotheses, the origins-of-life community tends to break into two main camps, each with its own implications for the exploration of ocean worlds.

The first camp argues that the best way to drive synthesis of larger molecules is through concentration and dehydration (drying) of a soup of smaller molecules. A great example of this, and a possible site for the origin of life, is tidal pools on the shores of our ancient ocean. One can imagine a cycle of wetting and drying as the ocean flows in and out with the tides (which on Earth were much larger and more rapid billions of years ago because our Moon was much closer).

As tide pools dry and water is lost to the atmosphere, molecules like amino acids have a stronger tendency to link together, forming small chains (peptides) that could have served as early proteins. Similarly, nucleic acids and sugars might have also combined more favorably through desiccation. The rocks and minerals within those tide pools could have aided in catalyzing these reactions, especially as the water evaporated and concentrated molecules. Finally, ultraviolet light from the Sun might also have served to catalyze some of these reactions, providing energy to zap molecules together. When the tides rose and more water was added, the cycle was refreshed, possibly adding additional raw materials to these chemical reactions.

Much of the organic matter assembled through these early, random reactions would have been useless "goo," but eventually—so the idea goes—self-replicating molecules would have emerged. Those molecules would have been the precursor to RNA, a critical chemical stepping stone toward the RNA-world. Once these successful pre-RNA molecules became incorporated into a compartment (proto-cell), the cell eventually flowed out into the ocean where it could continue to propagate. There life might have grown and reproduced as a soapy foam on the surface of our ancient ocean.

The second major camp in the origins-of-life community argues that another way to drive synthesis of large molecules, and perhaps close the "gap," is through chemical reactions around active regions of Earth's seafloor. Hydrothermal vents, such as those we visited in chapter 1, win the prize for everybody's favorite site for interesting deep ocean chemistry and possibly the origin of life. The prize is well-deserved. These systems are cauldrons for chemical reactions, and they've been stewing since the Earth gave birth to an ocean.

On Earth today vents are typically surrounded by beautiful and bizarre ecosystems with microbes and large creatures feeding on the chemistry of the vents, and on each other. On our young Earth, billions of years ago, hydrothermal vents might have provided abundant sites for the keystones for the origin of life to come together. Flowing up and through these early chimneys would have been compounds like methane, hydrogen, ammonia, and sulfide—all of which could have been synthesized into amino acids, sugars, and nucleic acids through interaction with the highly reactive minerals of the chimneys themselves. The chimney structures are large, porous, metal-rich geochemical environments that are like Mother Nature's own catalytic converters on the seafloor. As the hot fluids flow through the chimney, the tiny pore spaces within the chimney could have served as the lattice for making the compartments that eventually become cells.

One limiting factor for synthesis in these environments is that too much water could have been a bad thing. The hydrothermal vents are in the ocean, and thus anything that gets made in the vents might be rapidly diluted into the ocean at large. Contrast this with tide pools, where drying (desiccation) is part of the cycle. Desiccation can be a good thing when you want to link together small molecules like amino acids. By concentrating compounds, you can encourage them to bond together.

Furthermore, when two amino acids are joined together, a water molecule is made. If there is already a lot of water around, such as in the ocean, then this reaction is less favorable. Desiccation is therefore good. Sitting in a vent chimney at the bottom of the ocean—surrounded by water—might be bad.

Nevertheless, the general argument is that the chimneys act like big filters, and all the little pore spaces within the chimneys might help concentrate compounds locally, despite the diluting tendency of the ocean itself.

These two scenarios each have considerable pros and cons. Indeed, both scenarios could be correct—life need not be constrained to one pathway for its origin. But each holds different, interesting implications for our search for life beyond Earth. The tide pool scenario requires continents. You need continents, or at least islands, to provide the shores on which an ancient ocean could lap into tidal pools. Furthermore, it requires a planet with an ocean on its surface. And that ocean needs to be in contact with an atmosphere through which the Sun's energetic rays can penetrate, helping to drive that early metabolism. This is all possible on worlds like Earth and Mars. Billions of years ago Mars likely had seashores and tide pools. Sunlight would have poured through its atmosphere and helped catalyze any number of reactions. If the tide pool theory is correct, then Earth and Mars should both have life.

But the ocean worlds of the outer solar system lack continents and tide pools. Ice shells cut off any light from the Sun. The tide pool scenario simply does not work on ocean worlds. If that is the only way for life to originate, then these worlds should not have life. It would not be possible for worlds like Europa and Enceladus to be inhabited with life that arose within its oceans.

Meanwhile, for all the challenges that vents immersed in a young ocean might face from dilution, vent environments would have been easy to come by on early Earth, a young Mars, and even within the ocean worlds of the outer solar system. The deep ocean on Earth would have been partially protected from the chaos occurring on our young planet's surface. Impacts from asteroids and comets would have regularly wreaked havoc on early Earth, perhaps even occasionally boiling off much of the ocean in the early days. The deepest regions of our early ocean might have provided the only refuge for life during this period. On Mars and the ocean worlds, hydrothermal vents would have also served as sites of deep refuge for the origin of life. While Mars has long since lost its ocean, hydrothermal vents could be churning away today on the seafloors of worlds like Europa and Enceladus.

Returning to the beginning of this chapter, part of what motivates me and excites me about these two viable hypotheses for life's origins is that we can actually begin to test them through our exploration of ocean worlds. Within our own solar system, we have a diversity of worlds with a diversity of environments, many of which could be conducive to the origin of life.

If the origin of life is contingent on tide pools baking under a hot star, then throughout the history of the solar system, Earth, Mars, and even Venus might all have given rise to life. In this scenario, the alien oceans of the outer system would be devoid of life (barring the very small chance of transfer of life from another world). If we explore Europa, Enceladus, and Titan and fail to find even a whiff of life, then I think it means that the origin of life requires continents; it needs dry rocks and tide pools, and the cooking of a star from above. In that scenario, Earth-like planets are where to find life.

If, however, hydrothermal vents are preferred, then the deep oceans within the ice-covered moons might be teeming with life. Europa, Enceladus, and Titan may have given rise to life and may be sustaining life through to the present epoch. Perhaps all the worlds in our solar system that host, or have hosted, hydrothermal vents at some point in their history, also hosted life. From an ancient ocean on Venus to a bizarre chemical blend of liquids within Pluto, if the chemistry for life's origins is common, then life itself may also be common.

But it may well be that no world other than our own home, Earth, harbors life.

Ocean worlds provide a test for our hypotheses about how life originates. Either way—if we find life on these worlds or not—we learn about how and where life originates and what might have happened billions of years ago here on Earth. We learn about who we are, and where we came from.

CHAPTER 10

ORIGINS IN AN ALIEN OCEAN

There is a place on Earth that could be a window back in time. A window to the conditions that may have given birth to life. A window not only into life's origins on Earth, but also possibly its origin in alien oceans.

If the origin of life is possible around a hydrothermal vent, then the magical landscape of the Lost City hydrothermal vents, one kilometer below the surface of the Atlantic Ocean, might be that window. Lost City contains the kind of hydrothermal vents that we think could exist deep within the ocean worlds of the outer solar system.

I had the opportunity to dive down to Lost City in the summer of 2003; it was one of the most awe-inspiring experiences of my life. Lost City is a field of hydrothermal vent towers, made of glistening white carbonate rock. Some of the towers rise well over 100 feet above the seafloor, giving the impression of ancient cathedrals dotting this beautiful landscape. The name is perfect for this site—it looks like a lost civilization, hiding away in the depths of our ocean.

From a distance, the carbonate has a texture that reminds me of wax candles with solidified drips along the side. The work of the Spanish architect Antoni Gaudí, and in particular his magnificent Sagrada Familia cathedral, is the closest thing I can think of that resembles the tall towers of Lost City. It is a truly magnificent environment. And it's at the bottom of the ocean.

On this particular occasion, we were doing a dive that involved two relatively new submersibles called the *Deep Rovers*. This dive was part of James Cameron's expedition in 2003, and we had a total of four submersibles making simultaneous dives. Never before had four submersibles been deployed at the same time, at the same site. It was a bold and risky plan. The two Russian *Mir* submersibles were reliable but not immune to problems. The two *Deep Rovers*—one of which had me as cargo—were not well-tested, especially at the depth of the Lost City vents. Lots of things could go wrong.

The *Deep Rover* is essentially a glass sphere that can seat two people. It provides an incredible experience in the ocean because you can look around in any direction—you've got a full field of view. Unlike more traditional submersibles, you're not constrained to one or two tiny portholes to see the ocean. In a *Deep Rover* you feel as though you are in an inverted aquarium—we've transported our air-rich environment from the surface into the ocean, and the fish can look in on us the same way we stare into an aquarium.

The *Deep Rover* I was in, with pilot Tym Caterson, was deployed first. This meant that we got to the bottom of the ocean before anyone else and had some time to collect samples and explore before the full coordinated effort with the other three submersibles.

Tym and I found a nice spot on the bottom of the ocean, with one of the carbonate chimneys towering above us. After making a few ballast adjustments and acoustically communicating with the ship above to let them know we were safely on the bottom, Tym let me take over the controls. I used the robotic arm to grab a few rocks at the base of the chimneys. I tried to work quickly because I knew that as soon as the rest of the team got to the bottom, we would have to move to a new spot and begin the coordinated exploration.

When we first called up to the surface, we got word that the second *Deep Rover*—with Cameron and pilot Paul McAffrey—was experiencing some problems. They had put the sub into the water and would get back to us soon on their progress.

With that information, Tym and I were happy to continue collecting samples and exploring the bottom. At one point, as Tym was teaching

me how to control the sub, I turned the craft to the left, and to our astonishment, a huge floating creature drifted into view. It was right in front
of us, and it was magical. It was a cnidarian (the phylum that includes
jellyfish) spanning nearly 2 meters (6 feet) in diameter. It undulated
through the water like a pulsating hot-air balloon, only a few feet from
us. Its thin, translucent body made it possible to see the structure that
made the umbrella of its form. We could see the central pod of its mouth
and guts, its filter for food as the water passed through. Some of that food
was almost certainly microbes that had been grown by the chimney farm
of the Lost City vents. This jelly was like an aquatic wildebeest, grazing
on the food that formed the base of the food chain, food made possible
by the vents.[1]

With our jaws dropped at the sight of this magnificent creature, Tym
and I called to the other sub, imploring them to get down to the bottom
as soon as they could. Cameron had an array of lights and cameras on
his sub. We knew he would want to see this and capture this beautiful
creature on film.

But their problems persisted. We got word that they had a leak. A leak
in their sub. Generally speaking, one of the last things you want to hear
from your friends in a submersible is that their sub has a leak.

Nevertheless, they continued to descend. Counterintuitively, it was
not a crazy plan. To enter these particular subs, you climbed in through
a hole in the bottom, and the hatch was pushed up an into the hole, like
a cork in an upside-down wine bottle. Cameron and Paul reasoned that
as they dove deeper and the pressure increased, the cork would get
pushed tighter and tighter into the hole, and eventually the leaking would
stop. Sound logic, no doubt about that.

By the time they got to the bottom, the sub was still intact, but there
was a sizable pool of water on the floor. The leak appeared to have slowed,
and maybe even stopped, but now they were worried about short-
circuiting the sub's electronics. One of their thrusters had already quit,
and it was only a matter of time until more things would fail.

Literally under pressure, they managed to find us. Cameron captured
the cnidarian on film, and to this day it is the most alien-looking creature I've ever seen. For the rest of our dive, we swam around the majestic

towers, taking in these cathedrals of the sea. Were it not for the sight of that large creature, I could have easily convinced myself that we had traveled back in time a few billion years. The ethereal landscape of Lost City also transported my imagination to the distant depths of alien oceans. Europa and Enceladus, in particular, could have seafloors dotted with forests of chimneys comparable to those just beyond the glass of my submersible.

LOST CITY FOUND

Lost City was discovered in December 2000 on an expedition led by oceanographer Deborah (Deb) Kelley. It is a discovery that has had wide-reaching implications—from how our ocean and biosphere works, to how life might arise within oceans beyond Earth. What Kelley and her team discovered was a hydrothermal vent system unlike any seen before.

Hydrothermal vents were first discovered in 1977, and prior to 2000 every vent discovered operated under what I'll call the "blow torch" principle: the hydrothermal vents are driven by a heat source beneath the crust of our ocean. Like a blow torch heating a pot of water, the crust of the seafloor is heated by molten rocks rising from the interior of the Earth. This heating then creates active geochemical environments in the ocean—think volcanoes and hot springs at the bottom of the ocean.

The chemically rich fluids that flow out of these vents are what enable microbes to survive. Instead of drawing energy from the Sun via photosynthesis, they draw energy from the chemicals via chemosynthesis. These chemosynthetic microbes serve as the base of the food chain. Chemosynthesis makes life possible in these dark depths of our ocean.

At Lost City, however, the hydrothermal vents are not so much "blow torch" as they are "hand warmer." The heat comes not from molten rocks rising up through the Earth but from a chemical reaction that occurs when ocean water mixes with some of the Earth's deepest, heaviest, metal-rich rocks in the seafloor. The chemical reaction is exothermic, i.e., it creates heat, and it also produces a rich array of gases and minerals. The reaction process is called serpentinization because many of the rocks that are produced have a greenish, scaly appearance and texture, like a snake. One of the minerals made is called lizardite because of its resemblance to the skin of a lizard.

To give you a sense of serpentinization as an exothermic reaction, if you have ever felt one of those chemical hand warmers on a cold winter day, you have experienced the heat of an exothermic reaction. Exothermic reactions create heat as various compounds and minerals mix together. Hand warmers work by bringing together oxygen in the air with a very reactive form of iron powder in the bag. This is why the bags are packaged in a porous cloth, which is stored and sold in an airtight plastic package. The plastic packaging prevents oxygen from reaching the hand warmer until you want to use it. Once open, oxygen oxidizes the iron, essentially turning it to a form of rust, and heat is created as a by-product. In the process, your hands get warm.

The exothermic reaction of serpentinization is a more extreme and exotic version of rusting. Metal-rich rocks from the deep interior of the earth are highly reduced, which means they have lots of spare electrons they are looking to deal away in useful chemical transactions. However, everything around them deep below the seafloor is also reduced, and there are no transactions to be had; there are no oxidants available to accept their extra electrons.

But if some of this deep mantle rock gets pushed upward and becomes part of the seafloor, then these reduced rocks, such as peridotite, can mix with ocean water. Peridotite rocks are rich with the minerals olivine, $(Mg, Fe)_2SiO_4$, and pyroxene, $(Mg, Fe)SiO_3$. The iron and magnesium in these minerals are teeming with swarms of electrons. Even though we do not typically think of water as very reactive, it is when it combines with minerals desperate to share electrons.

Compared to the "blow torch" hydrothermal vents that billow out superheated water (over 400 °C) and minerals, giving the appearance of black smoke akin to a factory chimney, the Lost City system is very slow and serene. The towers do not billow out mineral-rich black smoke; instead the moderate-temperature (70–100 °C) fluids from serpentinization shimmer out of the carbonate chimneys, creating an ethereal look. Staring at them you might first think that your eyes are out of focus, but upon closer inspection you realize that the blurriness is from the refraction of light through the warm fluids emanating from the vents.

Why is Lost City so significant?

First, serpentinizing vent systems could be ubiquitous. They do not require the extensive plate tectonics and internal heating that drives the hot hydrothermal vents. Serpentinizing vents could arise whenever, and wherever, rocks like peridotite come into contact with water. This could have happened throughout the history of planet Earth, and it could happen on many worlds beyond Earth—from Mars to Europa and Enceladus. Serpentinizing systems could be very common. All that is needed is a cracking seafloor that allows water to percolate into the rocks. We do not know what it takes for plate tectonics to arise on a world, but it's a safe bet that cracking can occur in the seafloors of alien oceans.

Serpentinizing vents may also be responsible for generating life's first meal. The by-products of serpentinization include molecular hydrogen (H_2). In Earth's ocean water, both today and in the earliest days of our planet's history, there is and would have been plenty of dissolved carbon dioxide, and these two compounds create a reductant–oxidant pair that some microbes love. They eat hydrogen and carbon dioxide, and the waste they produce is methane and water. They're called methanogenic microbes because they generate methane.[2] Going back to the battery analogy from the previous chapter, hydrogen is the negative terminal and carbon dioxide is the positive terminal of the biochemical battery. This reaction is perhaps the smallest "battery" that life uses. It provides just enough power to drive life.

How small is this biochemical battery and how much power does it require? A single methanogenic microbe requires a power of about 10^{-18} watts.[3] To put that into perspective, a 100-watt incandescent light bulb can fill a room. We humans run on about 100–200 watts, which is about the same amount of power needed for a typical laptop computer. A single methanogen requires one one-hundredth, of one-quintillionth, of the power we humans use.

Finally, there is another compelling attribute of the Lost City–type of hydrothermal system. Methanogenic microbes may represent a very old way of life. Eating hydrogen and carbon dioxide to make methane is perhaps the oldest metabolism of life on Earth. When we look at the tree of life on Earth, methanogenic microbes are very deeply rooted; meaning that in some ways they carry the traits of the last universal common ancestor. In other words, if microbes, like humans, could carry on family traditions, these microbes appear to be carrying on one of the oldest traditions rooted in the tree of life. By understanding this organism, we may perhaps gain insights into the first life forms on Earth. Given that these microbes are found in abundance at sites like Lost City, perhaps the origin of life itself may be closely tied to these hydrothermal vents that support these methanogenic microbes.

ORIGINS FROM THE BOTTOM UP

In the last chapter we examined some of the requirements, constraints, and theories for the origin of life. One theory that I find very compelling suggests that life originated near serpentinizing hydrothermal vents, such as those found at Lost City.

Dr. Mike Russell, of NASA's Jet Propulsion Laboratory, is one of the founders of this theory, and he is strongly motivated by correctly framing the question of "life." When thinking about the origin of life, he argues, it does not make sense to ask what life is but rather what life does.

Why? Because if we ask, "What does life do?" the answer is that it alleviates chemical disequilibrium in the environment, completing the circuit so the battery can run down. As Russell sees it, the most basic "circuit" for life is to use hydrogen to transform carbon dioxide into some form of carbon compound, such as methane. Translating this into more conventional language, Russell (and numerous colleagues in this field) consider life's first metabolism—life's first lunch—to be the most critical piece of the puzzle: first figure out what life does, and how it did it. For Russell, it's about eating carbon dioxide and hydrogen, while exhaling methane.

I find theories like Russell's compelling because it places a premium on the energetics of the universe—it closes the gap we discussed in chapter 9. Interestingly, along with being a source to drive metabolism, vent towers such as those found at Lost City may also have provided a mineral matrix through which the chemically rich fluids flowed, and in which reactions were constrained. Part of the logic of the hydrothermal vent theories for the origin of life is that the porous minerals of the towers provide cell-like compartments.

Decades ago, the geologist Graham Cairns-Smith hypothesized that the mineral structures of the vents could have been the scaffolding that enabled the organic structure of life to come together, just like a building requires a scaffolding during construction. As with a construction site, once the building is complete, the scaffolding is no longer needed. Mineral surfaces can help catalyze certain chemical reactions and provide a place for organic material to adhere to. Perhaps, the porous minerals at the vents provided the first structures for life and later gave way to the self-sufficient organic cells that we have today.

Another attribute of the serpentinizing vents that may have helped life get its start is the pH and temperature of the fluids. By not being too hot (approximately 100 °C as opposed to more than 400 °C of the black smokers), the serpentinizing fluids do not cook any synthesized compounds. This is important because the bigger molecules get, the easier it is for them to be destroyed by high temperatures.

The alkalinity of the vent water at Lost City is quite high, with a pH value between 10 and 12. The surrounding seawater has a pH considerably lower. In Earth's modern ocean, the pH is nearly 7, but on early Earth it might have been as low as 5. This matters because pH is a measure of the number of free protons flowing around in the fluid.[4] These protons are in the form of H^+, which is simply a hydrogen atom that has lost its electron. A fluid with a pH of 10–12 has far fewer protons than a fluid with a pH of 5–7.

Consequently, within the walls of the vents, a proton gradient is set up across the chimney, running from the outside inward. The excess of protons in the seawater want to move into the water within the vents, which has a lower concentration of protons. This proton gradient could also be another piece of the puzzle. Life and cell membranes work in part through the flow of protons across membranes. The proton gradients in biology are part of how the circuit of life works. Could it be, then, that the difference in pH found at Lost City is a clue to the early workings of life? I think it may.

While Russell and many other colleagues have been working in their labs to simulate various pathways for the origin of life on Earth, Mother Nature has been slowly cooking away her own experiments elsewhere in our solar system. From a young, possibly watery Venus, to a warm and wet Mars, to the many ocean worlds of the outer solar system, the experiment of life's origins may have been conducted many times in many different ways throughout our solar system. Serpentinization, for example, could have been—or could be—a very common process on many of these worlds. If it is, perhaps it has laid the geochemical foundations for life throughout our solar system, and beyond.

One place where we think we already have pretty good evidence of active serpentinization is Enceladus. In chapter 6 we examined the chemistry of Enceladus's plumes in detail. Not only does that chemistry point to hydrothermal vents, it also specifically implicates serpentinization as the driving mechanism for those vents. The silica, methane, carbon dioxide, and hydrogen in the plumes point toward a low temperature, low pH, serpentinizing seafloor deep within Enceladus. Lost City here on Earth may have a geochemical cousin churning away in the Saturnian system. Given the low gravity of Enceladus, nearly one one-hundredth of the Earth's, I wonder if there might be vast carbonate chimneys, kilometers high, rising above the seafloor like geological sky scrapers, reaching toward the ice shell.

The origin of life is the critical first step from habitable to inhabited. It could well be *the* bottleneck that limits the abundance and distribution of life in the universe. Conversely, life's origins may be a straightforward geochemical inevitability, one that we scientists have just not yet cracked in the lab. Thankfully, we can explore the alien oceans in our solar system to help answer these questions. If life can and does emerge within alien oceans, the next step is sustaining that life through time and building a biosphere.

CHAPTER 11

BUILDING AN OCEAN WORLD BIOSPHERE

In the previous chapters we explored the energetics of life and looked at the specific case of life in hydrothermal vents. There, and in many places around planet Earth, the methanogenic microbes combine hydrogen (H_2), as the reductant, with carbon dioxide (CO_2), as the oxidant, to produce methane. This metabolic "circuit" is very useful but also limiting. A lot of other compounds come out of hydrothermal vents, many of which could serve as the negative terminal (reductant) for life's energetic needs. But only a few of those compounds could serve as the positive terminal (oxidant) that life needs. Hydrothermal vents are a great negative terminal for life but a poor positive terminal. More broadly, if we think at the planetary scale, the metal-rich interior of a planet makes for a good negative terminal, but a positive terminal can be hard to find.

The availability of oxidants to couple with reductants could be a limiting factor for any biosphere within an ice-covered ocean world. As detailed in previous chapters, the dissipation of tidal energy could lead to active seafloors capable of supporting hydrothermal activity. At Enceladus, we think that the *Cassini* spacecraft might have even tasted some of that chemistry from a serpentinizing seafloor. With seafloor activity comes the prospect of a wealth of reductants.

That is great news for life; but if there is no reliable source of oxidants, then biology could be a short chapter in the story of these worlds. If, for example, carbon dioxide coming from the seafloor is the only oxidant available, then life may originate, but it would be energetically limited—there might not be enough "food" to fill its ocean with life. In addition, if geologic activity on the seafloor died down and carbon dioxide diminished, biology would follow suit. No energy to power life means no life.

The story of life on Earth tells the tale of oxidants in spades. Microbial life on Earth may have arisen some 3.8 billion years ago, but oxidants were likely a limiting factor in the energetics for life. For over 3 billion years, life was limited by an inability to find a good positive terminal. Oxidants were hard to find, but reductants were easy—planet Earth was belching them out. The available oxidants were sufficient for microbial life, but nothing more.

Our atmosphere at the time was predominantly nitrogen, as it is today. But unlike the atmosphere of today, which consists of 21% oxygen, it had only a trace amount of oxygen. The oxidants available in the atmosphere and ocean were carbon dioxide, sulfur dioxide, sulfate, and a whiff of oxygen. Many of these oxidants were created in the upper atmosphere by ultraviolet light from the Sun bombarding compounds and turning them into oxidants. The oxidants were not nearly as abundant as they are today, and the ones that were available could only power small "circuits" for life. The available circuit pathways for harnessing the power needed for life were quite limited.

As a result, biology was so limited that it turned to space—*outer space*—to get over the oxidant hump. Our planet could not do the job, but the Sun could. Evolution eventually discovered the energetic niche and advantages of photosynthesis.

Significant debate exists within the scientific community as to when, exactly, photosynthesizing microbes (cyanobacteria) appear on the scene, but all the evidence indicates that photosynthesis[1] evolved as a metabolic pathway prior to 2.5 billion years ago.[2]

Before the innovation of photosynthesis, carbon dioxide was a run-of-the-mill molecule. Sure, it was popular, but no microbe in its right mind thought carbon dioxide was that great. When evolution stumbled on photosynthesis, however, the entire circuitry of life on Earth began to change.

By harnessing the energy of photons from the Sun, cyanobacteria were able to reduce carbon dioxide to sugars and exhale molecular oxygen (O_2). In other words, cyanobacteria figured out how to take carbon dioxide out of the ocean and atmosphere, extract the carbon atom, and turn it into a molecule that was useful for building the structures of life. The carbon atom was incorporated, or assimilated, into the structure of the organism. Doing that takes a lot of energy—a powerful circuit—and the Sun certainly had plenty of energy to give.

The brilliance of photosynthesis is that the Sun's photons are used to bump up the energy of electrons in carbon dioxide and water such that the carbon can be connected to other carbon atoms in a sugar molecule. It is a miraculous reaction pathway, and it is a testament to the trial and error of Darwinian selection. Once cyanobacteria appeared on the scene and began to photosynthesize, oxygen started pumping into our primordial atmosphere.

The real reason I love photosynthesis—and the reason you love photosynthesis—is because cyanobacteria and plants *exhale molecular oxygen* (O_2). The waste produced by photosynthesizing organisms is oxygen. Molecular oxygen is the most glorious of oxidants; it's like a universal positive terminal that also increases the voltage on your battery. I don't think there is any reductant that can resist the chemical temptation of giving oxygen an electron. Along with gases like hydrogen, methane, and sulfide, oxygen also loves to react with metals of all shapes and sizes. From iron to uranium, oxygen exerts its power across the landscape of elements and minerals.

Initially, to the best of our knowledge, there were no life forms—including microbes—available to use the oxygen made by photosynthesis. Much of the oxygen went into *rusting* planet Earth and our ocean. Iron that was dissolved in our young, acidic ocean began to combine with oxygen, leading to minerals that precipitated out and accumulated on the seafloor. Cyanobacteria continued to pump oxygen into the atmosphere; and once the ocean and continents could rust no more, the oxygen began accumulating in the atmosphere.

Gradually, the concentration of oxygen in the atmosphere increased: 1%, 2%, . . . , 10%, 15%. These changes started accelerating around

850 million years ago, and by about 600 million years ago, oxygen reached a concentration of 15–20% in our atmosphere. The large *negative* battery terminal within our planet finally had a large *positive* terminal in the atmosphere to match its potential. Oxygen was ready and waiting for the energetically innovative, at least by Darwinian standards.

Meanwhile, biology continued to evolve, mutation by little mutation, descendant by slightly modified descendant. Evolution eventually stumbled across the chemical machinery needed to use oxygen. Some of the microbes that figured out how to breathe oxygen merged with other microbes, giving rise to organelles like mitochondria. The mitochondria in our cells today have their root in a free-living bacterium that was good at using oxygen. Its skill was symbiotically co-opted and incorporated into other microbes. Although evolution is often framed as the survival of the fittest, Darwinian selection is also the story of symbiotic acquisitions and mergers.

As aerobic (oxygen-using) life continued to adapt and change, two friendly microbes (maybe it was several microbes, maybe it was millions) rubbed together and decided to cooperate. With the added energy from breathing oxygen, more complex life forms became possible. About 650 million years ago, microbes teamed up and gave rise to the first multicellular organisms. With oxygen as an energy source, teams of microbes could now work together to combine oxygen and organics in a very energetically advantageous reaction. Multicellular life using aerobic respiration was born.

Few evolutionary adaptations would change the face of the Earth as dramatically as the transitions to photosynthesis and then to multicellular life. The second of these transitions is typically referred to as the Cambrian explosion—a period in Earth history when the availability of oxygen appears to have led to a wide variety of bizarre and beautiful creatures, all working to breathe in oxygen and use it as the powerful positive terminal in their biochemical circuitry.

Oxygen in our atmosphere was critical for microbes and macrofauna (animals) on Earth. Oxygen also combined with other elements, such as sulfur, to create an additional menu of oxidants that microbes could eat. Most of the oxidants used by microbes at the deep-sea hydrothermal vents have some connection to oxygen produced by photosynthesis. Microbes at the hydrothermal vents that use sulfate are, in part, dependent on photosynthesis because photosynthesis generates the oxygen that is then transformed into sulfate within our ocean. Whether you are a microbe or a mammal, life is better when oxidants and reductants are plentiful.

For the past several hundred million years, oxygen has made up 21% of our atmosphere. If microbes and plants started growing more abundant and increased the level of oxygen, then forest fires would potentially run wild, burning down much of the oxygen-producing capability. Similarly, if animals were to take over and start consuming too much oxygen, then oxygen levels would drop and animals would no longer be able to survive. Plants, animals, and the atmosphere have a complex, self-regulating interaction that modulates the balance of oxidants and reductants.

As a short aside, this idea that the biology of a planet works, unintentionally, but through natural selection, to influence the geologic and atmospheric state of the planet so as to be more suitable for life is known as the Gaia hypothesis. The plants, animals, and microbes of planet Earth might all work in a biological symphony that keeps the drumbeat of life pulsing. James Lovelock first put forth this concept in his pioneering book, *Gaia*, in 1979. When it comes to the habitability of our home planet, I think Lovelock might have been on to something, and we return to this concept later.

The availability of oxidants, and oxygen in particular, was clearly critical to building the biosphere of the Earth. A problem arises, however, for biospheres within ocean worlds. On Earth, we navigated the oxidant problem through photosynthesis. Microbes used photons from the Sun to transform carbon dioxide into oxygen, which then built up in our atmosphere for later use by a variety of organisms, large and small. But on ice-covered ocean worlds, photosynthesis is not likely to be a viable niche. The ice will likely prevent light from penetrating to the ocean below. It may be that the vast quantities of liquid water real estate in those oceans goes unused by life because the lack of oxidants cannot support large, global biospheres.

There is one compelling way around this problem, at least for Europa. Solving the oxidant problem brings us back to the largest "structure" in the solar system: Jupiter's magnetic field.

THE RADICALS OF RADIATION

To answer the question posed above—where to find oxidants to sustain life—we consider the specific example of Europa and its interaction with Jupiter's magnetic field. To the best of our knowledge, none of the other ocean worlds experiences quite the same interaction. But first we need to set the stage by taking a closer look at the magnetic fields of Earth and Jupiter.

Earth's magnetic field is an incredible, protective field for life on planet Earth. Like a fictional spaceship from *Star Wars* using a force field to deflect incoming dangers, Earth's magnetic field shields us from many of the small but abundant dangers careening through space.

Chief among these dangers are energetic atoms and electrons spewed out from our Sun. Like a spinning dancer whose dress flows in and out, up and down, against her partner's legs, the magnetic field of the Sun changes and flows out into the solar system, brushing up against the many planets that dance around her. As the field flows, so too do the energetic particles, creating what we call the solar wind. Without an atmosphere and magnetic field, the solar wind can directly bombard and irradiate the surfaces of planets, moons, and asteroids.

In the case of the Earth, our magnetic field redirects many of the particles along the lines of our magnetic field. Some particles head toward the north magnetic pole, and some head toward the south magnetic pole. As those particles funnel down along the magnetic field lines, they collide with atoms and molecules of nitrogen and oxygen in our atmosphere, creating the beautiful shimmering sheets of green and red that we see as the Northern (and Southern) Lights. Thankfully, Earth's atmosphere and magnetic field help shield us from the worst of the radiation coming from the cosmos, redirecting it into the beautiful auroral display.

Europa, however, is not as lucky. Europa does not have its own magnetic field. It resides deep within the influence of Jupiter's field. If we could see the magnetic field of Jupiter, it would be the largest object in the night sky. Although Jupiter's field partially shields everything within its domain from the solar wind and cosmic rays, it picks up a storm of radiation of its own. Ions and electrons from Jupiter's upper atmosphere flow out into space along its field and create a rain of radiation on the surfaces of Jupiter's many moons (and any robotic spacecraft).

Adding to this is Io. For all of its volcanic beauty, with each eruption Io exhales a soup of sulfur and other compounds, which gets ionized and stirred into the mix of Jupiter's magnetic field. The reality of this radiation is easily observed in some of the images sent back by the *Galileo* spacecraft. Dotted throughout the images are little white pixels, each indicating a radiation hit that occurred on the detector while the image was being taken. With this in mind, it's no surprise that the surface of Europa is not a great place for humans to visit. Even robots, unconstrained by the limits of human frailty, are challenged by Jupiter's realm.

The surface of Europa is bathed in radiation. It's like a hail storm of energetic electrons and ions pouring down on Europa's surface, bombarding the ice and whatever else is there. Think of the occasional news stories you hear about large bursts of solar wind (also known as solar storms) wiping out satellites and electronics on Earth. Now imagine that you are floating high up in our atmosphere during one of those storms. That is what standing on the surface of Europa would be like. You don't want to stand on the surface of Europa; it would be the last place you stand. Sure, the view would be spectacular—Jupiter overhead, ice cliffs shimmering in the distance—but your chance to enjoy the view would be cut short by your exposure to the radiation.

In a curious twist of astronomical fate, the same storm of radiation that makes the surface of Europa harsh may also serve to feed life within Europa's ocean. If there is an ecosystem in the ocean beneath Europa's ice, the radiation bombarding the surface may be an important part of the chemical cycles that sustain life. The radiation may be the key to building an ocean world biosphere.

Here is the simplest description of what happens: when Europa's ice is irradiated, oxidants are formed. The radiation on Europa's surface creates the positive terminal of the biochemical battery.

Think about a water molecule (H_2O). When it is hit by an energetic electron or ion several different things can happen. The primary result is that the hydrogen and oxygen are split apart. In one of the most probable scenarios, H_2O breaks into H and OH; that is, a lone hydrogen atom and the oxygen-hydrogen pair (called the hydroxyl radical). "Radical" is a term used in chemistry to describe atoms, molecules, or ions that have incomplete valence shells. They want another electron to fill the slot in their outer shell, and for this reason they are highly reactive.

If ice is being irradiated, as occurs on the surface of Europa and to a much lesser extent on Enceladus, then the H and OH produced will occasionally combine with H and OH produced from a nearby water molecule that has also been irradiated. The result is H_2 and H_2O_2. The hydrogen, H_2, is very small and light and tends to drift off into space. The H_2O_2, also known as hydrogen peroxide—the same stuff you might have in your first aid kit—sticks around. Hydrogen peroxide accumulates in the ice, and occasionally it too gets hit by an energetic particle. When hydrogen peroxide is irradiated, a number of different chemical pathways arise; but one outcome is that the molecule splits and the H_2 escapes, leaving molecular oxygen (O_2) behind, in the ice. Irradiation of oxygen can then even lead to the production of ozone, O_3, in ice.

Another piece of the puzzle is that if there are other compounds in the ice, such as sulfur and carbon, then the irradiation of the ice causes those elements and molecules to be processed into oxidants such as sulfate, carbon dioxide, and carbonic acid. Sulfur on Europa comes from the volcanoes of Io, and we know that at least some carbon must be transported in from the solar wind.

The measurements made by the *Galileo* spacecraft were critical to revealing this curious radiation chemistry. When *Galileo* flew by Europa in the late 1990s, an absorption was observed in the infrared spectrum near 3.5 microns. Bob Carlson and the *Galileo* NIMS team worked to track down the cause of this little absorption feature and discovered it was due to hydrogen peroxide on Europa's surface.

Not long after those measurements were published, John Spencer, of the Southwest Research Institute, and Wendy Calvin, of the University of Nevada, Reno, used a telescope at Lowell Observatory in Arizona and observed oxygen in the surface ice of Europa. I should point out that both oxygen and ozone had already been observed in the ice of Ganymede, using ground-based telescopes and the Hubble Space Telescope. The surface of Ganymede is also radiation processed, although to a lesser extent than Europa, and thus the observation of oxygen and ozone makes sense there too.

How much hydrogen peroxide and oxygen are on Europa's surface? Quite a bit. In some regions, the amount of peroxide is about 0.13% by number abundance relative to water—that is, for every 10,000 water molecules in Europa ice, there are 13 molecules of hydrogen peroxide. It does not sound like much, but for comparison, the hydrogen peroxide that you get at the pharmacy is a 3% solution by weight, which means that if you dilute it with about 1.5 gallons of water, you get the concentration of peroxide found on Europa's leading hemisphere.

As for the oxygen, the absorption features observed by Spencer and Calvin indicate that oxygen may be trapped in the ice at an abundance (number relative to water) of 1% or more. Taken together, the concentration of peroxide and oxygen in the ice of Europa is small; but if it gets transported to the ocean, it could then dissolve into the ocean water and become the positive terminal for energetic chemistry of life. In other words, on Europa, the radiation chemistry on the surface generates oxidants that could be delivered to the ocean below, and those oxidants might be very useful for any life within that ocean.

An ocean covered in ice, blanketed by darkness could still have oxygen in it. Could there be enough oxygen to support large, complex creatures? I think it may, but that all depends on the geology of Europa's ice shell and how efficiently the material on the irradiated surface ice makes it down to the ocean below.

A CRYSTALLINE CONVEYOR BELT?

When it comes to the thickness of Europa's ice shell, there is a lot of room for debate. At the heart of the debate is a simple question: Is Europa's ice shell thin or thick? If Europa was a pizza, this would be the New York pizza thin crust versus Chicago pizza deep dish debate. It's intense. As amazing as the gravity and magnetometer data were to help sleuth out the existence of the ocean, they didn't help much when it came to the exact thickness of the ice shell above. The answer to this question and to the question of how the ice cycles with the ocean below could have important consequences for any life within Europa's ocean.

First, it's worth specifying that "thin" is still fairly thick. I would classify a thin ice shell as fewer than 10 km (6 miles) thick, and a thick ice shell as anything thicker than 10 km. As a point of comparison, the ice sheet of Antarctica is a maximum of 4 km thick. By Earth standards, Europa's ice is thick, no matter how you slice it.

But here is why this is important. If Europa's ice shell is thin, then it may be easier for all those radiolytically produced oxidants to reach the ocean below. The ice shell may crack and fracture from tidal forces breaking the ice apart. Material from the surface might then be mixed directly into the ocean, recharging it with a lot of biologically useful compounds. If, however, the ice shell is thick, and cycling of ice into the ocean is not as easy, then those oxidants on the surface might have a harder time reaching the ocean.

There are various nuances to the hypotheses and models for Europa's ice shell, and I won't go into detail here, but a key thing to appreciate is that we have very little data on the geology and dynamics of the ice shell. The *Voyager*, *Galileo*, *Cassini*, and *New Horizons* missions all returned various snapshots in time, and only the *Galileo* mission, which operated from the late 1990s to early 2000s, sent back a considerable dataset. We are thus left to argue the topic endlessly, until we get another spacecraft out there.

So what do we actually think we know?

We know that Europa's icy surface is very young, geologically speaking. On average, the ice shell of Europa is between 10 and 100 million years old. That may sound old, but for comparison, it is similar in age to Earth's seafloor crust, which is some of the youngest rocky material on Earth. The continents that make up the Earth are typically much older, from tens of millions of years old to several billion years old. The average age of the ice on Europa is comparable to the timeframe in which the dinosaurs were wiped out on Earth, which was about 65 million years ago.

How do we know Europa's surface is young, geologically speaking?

Recall the discussion in chapter 6 about Enceladus and how craters serve to indicate the age of Enceladus's icy surface. To the north, Enceladus is covered with craters; whereas to the south, there are essentially no craters. Craters are a clock that show the passage of time: more craters indicate that more time has passed. A lack of craters means that something has happened to erase any craters that were once there; the surface is fresh and young.

Europa's surface has not been thoroughly mapped at high resolution, but even with the limited imagery, it's fair to say that no place on Europa looks nearly as cratered as Enceladus's northern region. Much of Europa's surface looks comparable to Enceladus's southern region. Two of the scientists who have led the effort to clock the craters on Europa are Dr. Jeff Moore, of NASA Ames Research Center, and Dr. Beau Bierhaus, of Lockheed Martin. The analyses they led can be summarized as follows: If Europa's surface were older, we would not only expect to see more craters, we would also expect more large craters. The further back in time you go, the more large objects there were zipping around the solar system, ready to crash into something. The record of the craters, or lack thereof, tells us that Europa's icy surface is young. Whatever its thickness, the ice is somehow being recycled in a way that erases craters.

What Europa lacks in craters, it makes up for with other curious features in its ice.

When the first images of Europa started coming back from *Galileo* in the late 1990s, there was a lot of excitement about the appearance of its ice shell. Like a crystal ball shattered with a hammer, it has a cascade of fractures criss-crossing its surface. Previous images from the *Voyager* flybys showed a fractured surface, but now the extent and scale of those fractures could be fully appreciated: from the largest to the smallest scale, fractures were everywhere (Figure 11.1).

Along with the fractures, scientists also mapped out regions that looked like broken icebergs drifting apart from each other (Figure 11.2). These regions, named "chaos regions" because they look geologically chaotic, bore some interesting similarities to ice-covered regions on Earth and implied a relatively thin ice shell over a liquid water ocean on Europa.

FIGURE 11.1. Europa's icy shell is fractured across its entire surface from the effects of tides. Some of the largest fractures extend for thousands of kilometers. (Photograph from NASA/ JPL/University of Arizona)

How else could the ice on Europa be moving around and creating such a dramatic landscape? It had to be thin.

The planetary geologist Mike Carr led a team that concluded that the bizarre series of fractures and blocks of ice on Europa were indicative of an ice shell just a few kilometers thick. The rise and fall of tides cracked the ice directly, and heating of the ocean from the seafloor created warm regions that carried up through the ice shell, creating the chaos terrain.

As more images of the surface came down from *Galileo*, another idea emerged suggesting that some of the curious arc-shaped fractures cutting across Europa could be explained by the rise and fall of tides on Europa.[3] Tides had previously been implicated in the fracturing process, and now the math and computers models seemed to corroborate the idea. The idea, put forth by Greg Hoppa, Rick Greenberg, and team, showed that as the tides rise and fall, the stress pattern in the ice cuts an arc—or cycloid—across Europa's surface, potentially leaving behind fractures that

FIGURE 11.2. Parts of Europa's surface look similar to icy regions of Earth. Shown here are broken blocks of ice that look like icebergs floating in a sea. Several of the blocks have been overturned and cast stark shadows to the left (the Sun is to the right). However, Europa's surface is too cold (−280 °F) for floating and shifting ice, so warmer ice from below may be rising up and disturbing a brittle layer of ice on the surface. The area shown here is 50 km (30 miles) wide. (Photograph from NASA/JPL/University of Arizona)

match what was seen in the images. Such fractures could only occur, it was argued, if the ice shell was just a few kilometers thick. It was a beautiful pairing of mathematical modeling and observational evidence. In the earliest days of the *Galileo* mission, it seemed that the verdict was in: the cycloids, chaos, and fractures indicated that Europa's ice shell was thin.

But as time went on, the plot—and the ice—continued to thicken.

One key problem was that of cracking the ice. Although the rise and fall of tides was powerful, scientists like Dave Stevenson at California Institute of Technology argued that the stress and strain from tides would not be able to fully fracture an ice shell that was thicker than about four kilometers. That was not in and of itself a problem—perhaps the ice was even thinner than that—but it created a secondary problem. Could Europa provide enough heat to keep the ice shell that thin?

This problem of heat flow and balance is a central problem for understanding Europa, and, for that matter, all the ice-covered ocean worlds. Keep in mind that Europa's ice protects it from the cold vacuum of space—there is no atmosphere to blanket the surface. A thick ice shell is like a heavy winter jacket for the ocean below, insulating it from the cold of space. A thin ice shell is like a rain jacket on a freezing day—it may offer some protection, but you'll likely be cold. To stay warm, you better do some jumping jacks and create some internal heat. So too for the case of Europa with a thin shell. In order for the thin ice shell to work, Europa would need to be generating a lot of heat internally to counter the heat lost to space.

As powerful as the tidal heating might be, it was still a bit of a stretch to justify enough heat to maintain a very thin shell. One way to calculate the ice thickness is to measure Europa's surface temperature and then assume that heat is conducted from the ocean up through the ice. Thankfully, one of the instruments *Galileo* carried was a photopolarimeter, which is capable of measuring the thermal energy, or heat, coming off of a world.

From the *Galileo* measurements, we now know that Europa's surface is about −173 °C (100 K). The temperature of Europa's liquid water ocean is likely close to 0 °C (273 K). We know this from simple physics—even under the pressure of the ice shell, the temperature of liquid water in contact with an ice shell must be close to the freezing point (0 °C). Salt in the ocean might help suppress the freezing point, similar to what happens with seawater on Earth, but the temperature wouldn't get much below −10 °C, even for very salty oceans. When you take the surface temperature and the ocean temperature and plug them into the equation for conduction of heat through ice, you find that Europa's ice shell should be about 6 km thick.[4] That is likely too thick for the rise and fall of tides to crack all the way through the ice.

Counterintuitively, one way to rectify this discrepancy is to make the ice thicker. If the ice on Europa is thicker than about 10 km, then it likely starts to convect. In other words, the ice itself starts to flow, very slowly, like blobs in a lava lamp carrying heat from below up to the surface. Convection contrasts with conduction, which is the transfer of heat through a material by the microscopic jiggling of neighboring atoms and molecules.

You are likely familiar with convection happening with gases and liquids—from convection ovens to the flow of water in a pot on a stove—it is a way to transport heat by the movement of matter itself (i.e., the gas or liquid). Convection occurs in gases and liquids, but it also occurs in solid materials like rocks and ice; it's just a lot slower and thus less perceptible to us humans. Deep within Earth's mantle, rocks are slowly carrying heat up from the Earth's core as they convect upward and move, slowly but surely, up toward the Earth's crust. Once there, they release the heat they have carried.

On Europa, it may be that convection occurs within the ice shell, and the flow of ice carries heat from the ocean up to the surface. In this scenario, the thick convecting ice of Europa would have a thin cover of cold, brittle ice over its top. That cold brittle ice might be only a few kilometers thick and therefore could easily fracture by the stress and strain of tides.

One prediction of the thick, convecting ice model is that flows of warm ice from below (also known as "diapirs" in geology terms) would breach the surface ice of Europa and create patches of jumbled ice called "lenticulae," akin to "blisters" or "freckles" rising and popping on Europa's surface. This could be a good alternative explanation for the chaos terrain, and it also explains some of the smaller features, such as those seen in Figure 11.3. The chaos regions and lenticulae features on Europa could well be the surface expression of these convecting diapirs, rising up through a thick ice shell and breaking apart a crust of brittle ice on the surface.

The thick, convecting model for Europa's ice shell definitely wins the scientific popularity contest, and perhaps for good reason: it doesn't require anything special of Europa. The heating and fracturing problems are relatively well satisfied. That said, the cycloids and some of the higher temperatures measured on Europa still perhaps point to a thin ice shell.

How can these conflicting observations and models for the ice shell thickness be rectified? The answer may lie in the curious fact that few cracks cut across the chaos and lenticulae features. As Bob Pappalardo and colleagues have long argued, this clue may be an indication that the chaos and lenticulae formed after the many cracks formed.[5] Perhaps tidal heating was more intense millions of years ago and Europa's ice was much thinner; and tidal stresses cracked the thin shell and formed features like cycloids. Over time, the tidal heating decreased and the ice thickened,

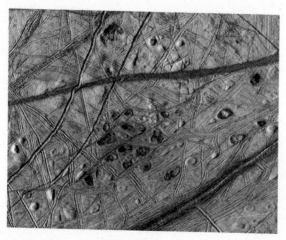

FIGURE 11.3. Tectonic activity marks the entire surface of Europa. The area in this image is 200 km (120 miles) across. The fractures have layered one on top of the other over time as the tides break and shift the ice. The dislocated fractures show once-contiguous lines that have been shifted, as perpendicular fractures have moved the ice in the opposite direction. The near-circular shapes—10 km (6 miles) in diameter—or "freckles" (lenticulae) may be the surface expression of warm ice rising up from below and beginning to breach the brittle ice on the surface. Some fractures and freckles are darker, perhaps indicating salts upwelling from the ocean. (Photograph from NASA/JPL/University of Arizona)

eventually leading to a thick, convecting ice shell that persists today. The young chaos and lenticulae features are from the rising diapirs that burp up onto the surface through that brittle ice layer.

This model is compelling as it fits with the ebb and flow of the Laplace resonance that we covered earlier in this book. The idea is that Europa may go through orbital cycles with Io and Ganymede that take tens to hundreds of millions of years, during which the tidal heating varies from moon to moon. It makes sense. As Hauke Hussman and colleagues have shown, the orbital dance that Io, Europa, and Ganymede do leads to changes in the eccentricity of their orbits, which then leads to changes in tidal heating. As the heating varies, the thickness of the ice shell changes accordingly.[6]

Could it be that millions of years ago Europa's ice shell was thin and cracking all over, but then it entered a period of lower tidal energy and the ice thickened, leading to a convecting, diapir-rich ice shell? It is a plausible model and explains well much of the evidence.

It is not a perfect model though. A few cracks do propagate through the chaos and lenticulae, and thus the timing of feature formation is not a perfectly clean story. Also, the chaos and lenticulae appear to be some of the most chemically rich, and perhaps saltiest, of regions. Diapirs will rise best if they are formed from clean, ductile ice that does not have much sea salt in it (ice with salt in it is heavier than clean ice and will thus not be buoyant).

So what is the answer—thick or thin? Interestingly, my colleague Cynthia Phillips at the Jet Propulsion Laboratory took a detailed look through all the images from the *Voyager* flyby and the *Galileo* mission with an eye toward any new fractures or other changes that might indicate current geologic activity.[7] Cynthia and colleagues found no obvious changes in the ice. Granted, the coverage of Europa's surface was low-resolution and relatively sparse, but it was still good to search. For now, the best we can do is to improve our computer models and use telescopes to collect data from afar.

To that end, recent work using the Hubble Space Telescope hints at the possibility that Europa has plumes jetting out from its surface. Such plumes would be direct evidence of current geologic activity and resurfacing of the ice shell. Two science teams, one led by Lorenz Roth of the Southwest Research Institute and the other by Bill Sparks of the Space Science Telescope Institute, have used complementary techniques; both indicate tantalizing evidence for plumes of water a few hundred kilometers high erupting out into space.[8] Those plumes could be connected to pockets of liquid water in the ice shell, or directly to the ocean itself. If the plumes are real, does that mean the ice shell is thin, at least in that region? If that is the case, could those oxidants on the surface be transported down to the ocean below? If processes like plumes are active today, then I think it bodes well for effective cycling of oxidants into the ocean, perhaps even if the ice shell is thick.

Beyond the plumes, I happen to think that Europa's ice shell is relatively thin, i.e., fewer than 10 km thick. Part of my rationale is that I think we may have some of our numbers wrong when it comes to the conductivity of heat through the ice shell. Most of the calculations and models to date assume solid water ice for Europa's shell. This is potentially problematic for two reasons.

First, the upper few centimeters to meters of Europa's surface might be more like snow than solid ice. This is important because snow is a much better insulator than solid ice. Think about an igloo. It can be nice and warm inside because snow is a very effective insulator. Meanwhile if you made an igloo from blocks of solid ice, you would probably have a pretty cold place to call home.

If Europa's surface has a layer of low-conductivity material like snow on its surface, then those calculations about heat transfer and the thickness of the ice could be wrong. With an insulating blanket of snow, Europa's ice shell could be thin (less than 5 km), and yet it would still look as cold as the measurements indicate (100 K).

Compare this to a frozen lake. If the lake is covered with solid ice, then the ice will continue to thicken because the cold air above will cause the water below to freeze. If, however, a snow storm rolls by and drops a foot of snow on top of the ice, then the freezing rate of the ice below will slow down. The ice and water are protected from the cold air by the insulating snow. The same basic principle applies to Europa. There the snow would not come from clouds but rather from plumes and other processes depositing fresh material onto the surface. I think our calculations are a bit off because we haven't correctly accounted for a fluffy, porous, snow-like surface.

The second issue that I think needs more consideration is the chemistry of the ice shell and how that affects the heat balance and the mechanics of fracturing. Europa's ice could be laden with salts, sulfuric acid, hydrogen peroxide, oxygen, chains of sulfur atoms—all sorts of miscellaneous compounds could be mixed in with the water ice. Adding these compounds into the ice has the potential to dramatically change the thermal conductivity and mechanical strength of the ice shell. The thermal conductivity could be greater than or less than that of pure ice, depending on exactly what is in the ice. Thus, when it comes to the ice thickness, it's hard to say exactly whether the net effect would be toward a thinner or thicker ice shell.

The mechanical strength, however, would almost certainly be decreased if the ice was salty and full of compounds other than water molecules. Some of those compounds would fill in small cracks and boundaries between ice grains, making it more difficult for the ice to form large, strong crystals. If the ice shell is easier to fracture because of these other compounds, then perhaps the tidal stresses could fracture a thicker shell. Maybe the ice shell is 8 km thick, but because it's full of salt and sulfuric acid, it is easier to fracture than a pure water ice shell.

We do not yet know the bulk properties and chemistry of the ice shell, but we do have a few clues from our spectroscopic interrogation of Europa's surface. These clues may help us better understand whether or not the oxidants in the ice make it to the ocean, and whether materials that are in the ocean make it to the surface above.

A WINDOW INTO THE OCEAN BELOW

What goes up must come down. And what goes down must come up.

All of the ice on Europa at some point came from the water of the ocean below. The ocean water may have jetted out directly from erupting plumes; it may have frozen into the ice and slowly floated up as a rising diapir; it may have sloshed up through various fractures and ice blocks; or it may have come up from some yet-to-be-understood geological mechanism.

Similarly, the ice and other compounds on the surface of Europa must, at some point, make their way back into the ocean. It's a simple but important statement, and stems from the following: the young age of Europa's surface; the chemistry we see on the surface; and the fact that if new material from the ocean below is creating the fresh surface of Europa, then the material that was on the surface must eventually be pushed back down into the ocean and recycled.

At some level, the ice shell must serve as a conveyor belt, bringing material from the surface down to the ocean. Similar to the rock cycle on Earth, where new rocks are made on the seafloor and in volcanoes while old rocks are covered and subducted, there must be some form of ice cycle on Europa that refreshes the ice we see.

Returning to the topic of radiation and the production of oxidants on Europa's surface, the geologic cycling of the ice is central to understanding whether or not the oxidants reach the ocean. If the oxygen, peroxide, sulfate, and other compounds are geologically transported to the ocean through the recycling of the ice, then the chemistry of the ocean might be able to sustain life for long periods of time.

In my opinion, one of the most compelling indications of ice cycling with the ocean below is the yellowish-reddish-brown material on Europa's surface, which appears—spectroscopically—to be salts. Those salts must be from the ocean, and thus the presence of the salts is evidence of delivery upward of material from the ocean. The salts are compelling evidence that the ice shell serves as a window into the chemistry of the ocean below.

If salts come up from the ocean, does that mean material on the surface makes it down? Perhaps.

I think it's likely because as new material is emplaced on Europa's surface, whatever had been there then gets mixed deeper into the ice and perhaps pushed down into the ocean. Furthermore, the ice on the surface cannot continue to accumulate salts without some of it being cycled back into the ocean. If it did not cycle back into the ocean, we would see nothing but salts on Europa's surface.

A few of the images of Europa's surface may tell the tale of ice moving from the surface into the depths. My colleagues Louise Prockter, of the Lunar and Planetary Institute, and Simon Kattenhorn, of the University of Alaska, identified places where they think the disappearance of fractures and cracks indicates that the ice is being pushed back down into the deeper parts of a thick ice shell, and possibly into the ocean itself. On Earth this process in the rock cycle is called subduction. For Europa and its icy shell, Procktor and Kattenhorn coined the termed "subsumption" to describe the icy version of this process.

Based on the geologic and chemical evidence described above, I think there is good reason to predict that the oxidants produced by the radiation on Europa's surface could make it into the ocean. The ice shell would thus be like the positive terminal in the biochemical battery. The chemistry of the ice could be feeding a deep ocean ecosystem.

At the most basic level, oxidants such as sulfate would be tasty food for any number of Europan microbes. Who knows what these microbes would look like and whether they would have DNA, but the sulfate from the ice could be combined with hydrogen or methane from the seafloor, thus providing a nice lunch for these alien creatures.

Could there be enough chemical energy to power even larger creatures? Might Europa's ocean have experienced anything similar to what happened on Earth, as life evolved from simple to more complex multicellular life? Is there enough oxygen in Europa's ocean to support creatures larger than microbes?

A number of years ago my colleagues and I calculated how much oxygen could be delivered to the ocean, assuming a handful of different geologic mechanisms. We learned that for most scenarios, the microbes would be fine. Microbes love to use sulfate as an oxidant. Even if only a fraction of the sulfate on Europa's surface gets delivered once every 100 million years, there is still enough sulfate to feed an ocean of microbes. Larger creatures, however, need an efficient delivery of the oxygen to keep them going.

If Europa's ice shell is thick today, then convection of ice, up and down, may be slower and less efficient than it was during periods of a thin ice shell. Were it to take 100 million years for the oxygen on the surface to make its way down to the ocean, then there is a chance that tiny, oxygen-using organisms could survive on this limited oxygen supply. Survival would be hard but not impossible, as the oxygen levels in Europa's ocean would be comparable to some of the most oxygen-starved regions in our ocean. In these areas, organisms like polychaete worms can survive, but larger creatures cannot.

For cases of a thin ice shell, where cycling of the surface ice might happen within tens of millions of years, it is relatively easy to deliver enough oxygen to the ocean such that dissolved oxygen in Europa's ocean is comparable to that of our ocean. In that case, Europa's ocean could have enough oxygen to support large creatures comparable to fish, squid, and octopi.

The answer, in theory, appears to be that Europa's ocean might be able to support larger, multicellular life. This is a profound conclusion: the irradiation of Europa's surface could lead to an ocean capable of supporting both single-celled microbial life and possibly even larger multicellular life. This does not, of course, tell us whether or not evolution would lead to the emergence of multicellular life. But at least the chemical energy from oxygen might be available to encourage evolution in that direction.

GAIA'S REDHEADED STEPCHILD

Gaia is a beautiful concept. It says that the biology of a planet actively influences parameters (such as the geology and the atmosphere) of that planet to be more suitable for life. I mentioned the Gaia hypothesis near the start of this chapter in the context of Earth's atmosphere. Other examples of parameters hypothesized to be regulated by Gaia include Earth's temperature and possibly the salinity of the ocean.

James Lovelock, who first suggested the Gaia hypothesis, was working at JPL when he was developing his ideas. Perhaps surprisingly, he did not begin his book *Gaia* with a polemic on environmentalism. Instead, he starts with a description of Mars and Venus.

Lovelock's research on other planets, and his curiosity about the atmospheres of Venus and Mars, led him to speculate how Earth could have sustained such a "perfect" atmosphere and temperature for so long. What was the "special" ingredient that made it possible for Earth to persist in such a habitable state?

The answer, he concluded, was perhaps life itself. And thus the Gaia hypothesis was born. Over the past many decades there have been debates and studies to examine whether or not the Gaia hypothesis is actually testable science. Either way, I find it to be a beautiful framework for considering planetary-scale ecosystems.

If the concept of Gaia applies to planet Earth, could it apply elsewhere? What might Gaia "look like" on other worlds? In particular, what would a biological Gaia look like on an ice-covered ocean world?

Central to this last question is the following: What would life want to regulate in order to make that world more habitable?

On worlds like Europa and Enceladus, I think the answer may be the ice shell. Life would want to control the ice shell, and the mixing of the ice shell with the ocean.

Why? Because, as we've learned in this chapter, the ice shell is a source of chemical energy that life can eat and use. The radiation that hits Europa's surface splits apart water molecules and other compounds on the surface and drives the production of hydrogen peroxide, oxygen, sulfate, and other oxidants. These molecules—and oxygen in particular—would be very important to any microbes and large complex organisms in the ocean below.

But these molecules are trapped on the surface. Unless there is a way for the geologic activity of the ice shell to transport those molecules down into the ocean, it's game over for the organisms that want to feed on that material. Life on Europa would want to control the geology of Europa's ice shell.

One key parameter that plays a role in all the ice shell and geology models is the grain size of the ice. What do I mean by "grain size"? Imagine a big sheet of ice. If you could pick apart that ice sheet, you would see that the ice crystals within have a variety of different sizes—those are the grain sizes.

On Earth, ice grains within glaciers and ice sheets can be microns to many centimeters in size. In general, the smaller the size of the grains, the easier it is for the ice to move and convect. Returning to Europa, small ice grains within the ice shell might enhance convection and cycling of ice with the ocean below. Life within the ocean might thrive if it could somehow help regulate the cycling of the ice into the ocean.

Interestingly, we see from microbes on Earth that they excrete a type of material broadly called extracellular polymeric substances (EPS). This EPS can be like an organic lubricant between the grains of ice, and it does in part serve to keep the grains of ice smaller than they would be if no life was present. In other words, based on what microbes on Earth are able to do, it is possible that alien organisms within an ocean on Europa could help control the grain size of ice, and in so doing, regulate the cycling of the ice with the ocean, bringing in those tasty oxidants from the surface above.

I enjoy the thought that over eons of evolution and the tidal dance of Europa around Jupiter, the yellow, red, and brown color on Europa's surface might be a clue to life in the ocean below. The salts and other material from the ocean might be efficiently upwelled to the surface by merit of life in the ocean releasing compounds that make the grains in the ice shell small and easy to convect. The ice shell of Europa might not only be a window into the ocean below, it could also contain the fingerprint of life within the ocean.

I love to imagine that the underside of the ice—where the ice meets the liquid water ocean—is teeming with microbes and possibly more complex organisms; creatures pitting into the ice to harvest oxygen and other compounds made by the radiation above and delivered through fractures and convection into the ocean.

Life on the underside of the ice on Europa could be like an inverted version of life we see lining the cracks of a sidewalk—those cracks are where life digs down and finds nutrients. On Europa, the various oxidants might be concentrated along the boundaries between ice crystals, forming veins trending upward from where the ice meets the water.

Microbes could line those veins in the ice while bizarre Europan octopi with tentacles mine the ice for nutrients, eating the microbes and the oxygen, while crawling along the underside of the ice. An entire alien ecosystem might exist, feeding off the chemical energy contained within the ice.

CHAPTER 12

THE OCTOPUS AND THE HAMMER

Imagine . . . never even thinking, "We are alone," simply because it has never occurred to you to think that there's any other way to be.

—Slartibartfast, *Life, the Universe, and Everything*[1]

In Douglas Adams's multivolume work, which includes *The Hitchhiker's Guide to the Galaxy*, his character Slartibartfast ponders the thought above as he looks out on the landscape of planet Krikkit. The inhabitants of Krikkit cannot see the stars because of a persistent thick layer of clouds preventing them from ever seeing the night sky. The civilization on this fictional world has no exposure to the stars above and thus no vision for what lies beyond. They do not even know to ask the question: Are we alone?

How would we humans evolve and think about our place in the universe if we could not see the stars? And how has this backdrop influenced our compulsion to explore? The night sky, the Sun, the Moon—they have defined our horizon. If you have no horizon, if you have no rising or setting Sun, will you still walk or sail to the edge of the Earth?

What if, for the past several million years of human evolution, instead of looking up at the night sky and seeing stars, we saw a solid shell of ice above us? What if we lived within a world like Europa, with a ceiling of ice in the sky? And what if we could not even see the ice because it's dark and no sunlight penetrates; instead we could only feel and hear the ice as it creaked and cracked. And what if that ice ceiling were full of nutrients to feed and sustain life?

Within the ocean of an ice-covered moon, would there be an evolutionary pressure to develop into an intelligent, tool-using species? If enough oxygen was available, how might life evolve (or not) from microbe to multicellular life to intelligent, tool-using life? What selection pressures could lead to technology?

To be perfectly clear, I am only speculating about the potential for complex life and civilizations to emerge on ice-covered ocean worlds. We are moving beyond the science of what we know into the realm of what might be possible.

At the highest level, this chapter is an exploration of what is *contingent* versus what is *convergent* when it comes to the evolution of intelligence and technology. Contingent developments in evolution are directly tied to, i.e., contingent on, specific attributes or events in an environment. For example, the sizes of a birds' wings are contingent on the thickness of our atmosphere and the gravity of the Earth. Similarly, the color of plants is contingent on the color of light coming from our Sun.

Convergent developments are more universal. Although environments may differ, these adaptations are ones on which biology converges because the solutions are so useful. Eyes, limbs, and skeletons in some form have all emerged numerous times in vastly different parts of the tree of life on Earth; evolution has converged multiple times on these adaptations.

The line between contingent and convergent adaptations can often be blurred, and that is part of why it is interesting to consider the case of intelligence and technology on alien worlds, and in particular, deep under the ice of an ocean world.

It may be that ocean worlds like Europa are ubiquitous throughout the universe and provide the most habitable real estate. If our solar system is any guide, ocean worlds may provide ten to a hundred times the volume of liquid water found on worlds like Earth, with oceans on their surfaces. What might this mean for the emergence of intelligent life throughout our solar system and beyond? Could intelligent life be swimming in oceans covered with ice, unaware of the universe beyond their attic of ice?

WHAT MAKES SENSE?

A fundamental set of conditions that influences intelligence is an animal's set of senses. Sight, sound, smell, taste, and touch provide the input signals for our perception of the world. They provide the information on which we base our decisions. These sensory modalities are the foundation for the evolution of human intelligence on planet Earth.

Consider then the senses of an organism in an ice-covered ocean, and what it might have as its sensory repertoire. Hearing, smell, taste, and touch all have advantages in our deep ocean, and these senses have all evolved independently many times on Earth, indicating a convergence that could be universal.[2]

Similar environmental conditions could exist in oceans beyond Earth, and thus these senses could provide a survival advantage and be selected for as evolution marches along. Sound propagates well in water, which makes hearing useful, and compounds large and small flow through ocean water with relative ease, making smell and taste advantageous. Lastly, touch is required for any organism moving in its environment. On Earth, the lateral line organ in many fish is used to sense vibrations, pressure gradients, and other movements of the surrounding water. It acts like a much superior version of human skin, able to sense even tiny vortices caused by nearby predators and prey.

It is reasonable to predict, therefore, that evolutionary convergence might arrive at similar sensory solutions. But would creatures in an ice-covered ocean—an ocean cut off from the Sun—develop eyes and sight? Could they see? Would they need to see?

Being able to sense and see predators and prey is an obvious advantage. So too would be the ability to see the hot spots of hydrothermal vents. Vents would potentially host ecosystems with other animals to eat, and thus finding vents means finding food. But on Earth, and in our ocean, sight works by collecting photons of light that have traveled from the Sun and bounced off the object we are seeing. At night, the Sun goes away and we can't see a thing. Within ocean worlds that have no source of photons from a star above, are there any other viable options for sight?

It turns out that hot rocks may be useful. Hot rocks, water, and minerals, such as those found at hydrothermal vents, emit a low flux of visible light photons as they radiate heat into the ocean. While most of the heat is in the form of infrared photons—which are of too low energy to drive photosynthesis—scientists have isolated a type of sulfur-eating bacteria from hydrothermal vents that may actually use the rare visible light photons to do photosynthesis.[3] There is even some evidence that crabs crawling around the chimneys may also be sensitive to visible light from the vents.[4] Clearly when eruptions are hot enough, and fresh lavas flow out onto the seafloor, both infrared and visible light are abundant. Those conditions, however, do not provide for stable, long-term environments.

Eyes to sense relatively stable hydrothermal vents would likely have to be sensitive to infrared light. One of the main problems with using sight to seek out hydrothermal vents, however, is that all the ocean water surrounding vents makes it impossible to see heat at large distances; the cold water covers up any heat signature.

This is part of why we do not yet have a full map of all the hydrothermal vents on Earth—the water of the ocean masks the heat signature. Even for visible light, the ocean water scatters photons and limits visibility to a few hundred meters.

To see an infrared hot spot like a hydrothermal vent, you would likely have to be close to it, typically within tens to hundreds of meters. Even then you would need large eyes and the ability to pick up the infrared heat signature of vents. Larger eyes mean a larger collecting area. Imagine an octopus with eyes the size of basketballs. Because the photons might be scarce, the eyes might not be great for seeing objects in detail, or in color for that matter, but they might give you a blurred view of sources of heat on the seafloor. To the eyes of an alien octopus, a hydrothermal vent might look like a bonfire at the bottom of the ocean, with blurred flames of hot water streaming up and out of warm chimneys.

Another possible evolutionary driver toward sight could be light coming from an organism itself, i.e., bioluminescence. In the deep ocean of the Earth, bioluminescence plays a large role in ecosystem dynamics and could, in theory, be useful within alien oceans.

But there is a bit of a chicken-and-egg problem. On Earth, the eyes that are needed to sense bioluminescence evolved up on the surface, where sunlight made seeing easy. As fish and other creatures evolved to survive in deeper, darker depths, eyes evolved to become more sensitive. Bioluminescence—for hunting, avoiding hunters, and mating—also became an effective strategy. But if deep ocean creatures did not have eyes to begin with, would bioluminescence ever evolve as a useful strategy?

Interestingly, the origin of bioluminescence may be connected to the rise of oxygen in Earth's atmosphere and ocean, a topic we covered in chapter 11. When complex life first arose on planet Earth, excess oxidants could have been a problem. As useful as oxygen is from an energy perspective, too much of a good thing can be bad. Think of all those "antioxidant" juices and diets that are meant to make us healthier.

One possible strategy for protecting against the problem of too much oxygen might have been the development of enzymes that help "burn" oxidants in a chemical reaction that produces a burst of light.[5] Instead of causing damage to important biomolecules, the enzymes harness the energy of the oxidant to make light. In this case, light is a waste product, almost like a spark released when connecting the terminals of a strong battery. The beauty of bioluminescence may have begun with this controlled burn of excess oxidants.

If microbes at the base of Europa's ice shell developed a similar survival strategy for dealing with too much oxygen, then they might luminesce. Bioluminescence might be a convergent solution to this problem. A larger organism feeding on those microbes could benefit from the ability to sense emitted light, thus leading to a selection pressure for sight.

The possibility also exists that the ice shells covering ocean worlds might vary in thickness over time, which could lead to ice thin enough (less than 500 m) for a faint whiff of sunlight to make it through. Consider the thickness of the ice shell of Europa, as we discussed in the last chapter. Early on in the history of the solar system, and perhaps periodically, Europa's ice shell might have been very thin. Throughout Europa's history and its tidal dance with Jupiter, Io, and Ganymede, there would have been periods of more intense tidal heating, during which the ice shell might have become very thin. There is a chance—perhaps small but nevertheless enticing—that those thin ice periods provided enough light to drive the evolution of eyes. Creatures living at the ice–water interface might use the light of the Sun to see, or even to photosynthesize. The evolutionary incentive toward sight could emerge.

Returning to the full suite of senses that might provide a survival advantage in an ice-covered ocean, the driver would be the ability to locate hydrothermal vents or to locate interesting chemicals streaming down from the ice shell. Consider the ways in which we search for hydrothermal vents with ships and robotic vehicles. Searching for plumes of hydrothermal vents is a bit like searching for smoke to find fire. On land, if you find smoke, you find fire. In our ocean, when you find a plume of chemically rich water full of particles, you can, if you're lucky, trace that plume back down to a field of hydrothermal vents. Intriguingly, on Enceladus and Europa—where the gravity is lower, the Coriolis forces are different, and the ocean is not stirred by wind—the upward motion of a plume could rise for tens of kilometers above the vents[6].

One of the top scientists with a knack for sniffing out hydrothermal vents in our ocean is my colleague Dr. Chris German of the Woods Hole Oceanographic Institution. For decades, Chris has been using the "where there's smoke, there's fire" principle to find hydrothermal vents, and he has successfully found vents all around the globe.

The instruments that Chris and his colleagues use are towed behind a ship as it moves from one spot to another. The line with the instruments is usually hauled up and down through the water, like a yo-yo, so measurements can be made at many different depths. Plumes in our ocean rise for a few hundred meters above the vents, at which point they spread out horizontally, like a cloud of smoke flattening into the shape of a large mushroom. Towing the instruments up and down improves the chances that they might dip through the large flattened region of a plume, which is much easier to find than the narrow "stalk" of the plume. This process of searching for the plumes is fittingly called "tow-yo-ing."

The measurements made include conductivity, to measure changes in the salinity of the water; temperature, to detect any warm spots; and turbidity, to look for particles in the water. The signature of a hydrothermal plume is an increase in these three parameters. Once you find that signature, it's time to send in the robotically operated vehicle to search for the vents up close.

From a biology perspective, the key senses an organism would need to track down hot, chemically rich hydrothermal vents would be a combination of taste, smell, touch, and sight. Add in the ability to sniff out methane, hydrogen, and hydrogen sulfide, and now you have the equivalent of a deep-ocean hound dog that could track the scent of hydrothermal vents over long distances.

This same set of senses could also be very useful for tracking down chemically rich regions at the base of ice shell. Salts, oxidants, and sulfur compounds might stream out of the ice, forming salty icicles called brinicles, that are like inverted hydrothermal vents. Instead of hot water, these structures send out super-cooled water that has not yet frozen into ice because the high salt content suppresses the freezing point. Channels of very salty, super-cooled water might run through a network of fractures in the ice shells of ocean worlds, eventually dumping out into the ocean at the base of the ice shell. The ability to sense and track cold, salty water might lead creatures to parts of the ice shell where oxygen and other useful compounds could be found.

And so, the senses of an organism in a complex ocean world might include hearing, smell, taste, touch, and even possibly sight. The five key senses we, as humans, have depended on—throughout our march from the beautiful plains of Africa as *Homo erectus* to the confines of an office cubicle as *Homo sapiens*—could also be a useful part of evolution within distant ocean worlds. These five senses have done wonders for our intelligence and ability to use tools. They might also motivate intelligence and tool use in ocean worlds.

The possibility also exists that other senses, which we as humans do not have, could be important to life within dark oceans.

Hearing is a good example. Along with hearing would likely come echolocation, similar to what dolphins and bats use. Since sound propagates well in water, the use of acoustics would be of high evolutionary value. Hearing could be elevated to a "super-sense," used for both communication and navigation.

Sensing electric and magnetic fields is another modality to consider. We humans do not, to the best of our knowledge, actively sense these fields, but plenty of creatures do. Sharks, rays, and many kinds of fish can sense electric fields in water—this is called electroreception. Numerous amphibians and even a few mammals, such as the duck-billed platypus and possibly dolphins, are also capable of electroreception. These animals can "see" the electrical fields created by objects in their surroundings; this information is useful both for hunting and navigation.

Electroreception only makes sense in liquid water environments. Salty seawater, and even freshwater, are conductive enough for electric fields to propagate. Air, however, is a huge resistor, making biological electroreception impractical for us land creatures.

The most common type of electroreception is passive, meaning that the shark or fish does not create the field it uses to sense things, rather it senses the weak fields created around it. Active electroreception, however, does exist, and it involves a separate organ used to generate and sense the fields.

The Mormyridae family of freshwater fish, native to Africa, are perhaps the best example of active electroreception.[7] The Mormyridae, sometimes called elephantfish because of their long snouts, have an organ near their tail that can be pulsed to create an electric field that flows out into their environment. Disturbances to that field can be caused by other organisms that are conductive, or by resistive materials like rocks. In some ways, active electroreception is similar to echolocation, but instead of using sound to map out the environment, electricity is used.

Intriguingly, active electroreception is also used for communication between fish, most likely for mating and defending territory. The ability of these fish to respond is so sophisticated that they can change communication frequencies if there is too much electromagnetic "noise" obscuring their signal. Simon Conway Morris has likened this to airplanes sending out their call signs and switching frequencies through a jamming avoidance response.[8]

Given that electric and magnetic fields propagate and change in many of the known ocean worlds, I like to speculate that perhaps electromagnetic sensing could provide a survival advantage. Within Europa or a similar ocean world, electroreception could be used for hunting and possibly for navigation. The changing electric currents in the ocean could provide a systematic pattern for knowing where in the ocean one might be—remember, they can't navigate by the stars or time of day! It might also be possible for some organisms to directly harness the electric currents within the ocean for power. They might capture the electricity directly, or use it to split water into hydrogen and oxygen, essentially creating a biological fuel cell.

The array of possible sensory systems available to deep ocean creatures does not, in my opinion, preclude the emergence of intelligent life within an ice-covered ocean world. The environment is complex and changing, meaning that intelligence has survival value.

The path toward the development of tools and technology, however, is less clear.

ON THE IMPORTANCE OF STUFF

Why do some creatures develop intelligence while others develop both intelligence and the ability to make and use tools? While many land creatures developed versatile fingers and hands, the most advanced arms and "fingers" in the sea are limited, for the most part, to the octopi. But at what point will an octopus finally use a hammer? With that set of eight amazing arms, it seems odd that the ocean has remained devoid of technological innovation. Why aren't there octopi that are Fortune 500 CEOs?

This is not to say that our friends in the ocean do not use tools. Plenty of examples exist. Sea creatures have developed ways for picking up rocks and shells, employing them for shelter or to crack open a meal. Dolphins have learned how to use their bubbles to herd fish for prey. Horns, stingers, ink, and bioluminescence are all incredible adaptations, but they are not standard tools—they are part of the organism. For the most part, the tools used in the ocean are relatively simple and they are not separate from the organism.

To be clear, the questions above have no obvious answers, and my goal is not to dive into the evolution of consciousness, but rather to provide a few insights into what I think may be limitations, or opportunities, for the emergence of intelligent, technologically advanced life within ocean worlds. This is all highly speculative and immensely fun to ponder.

As an initial constraint on the survival advantage of tool use, let's consider mobility. I think mobility is a particularly relevant attribute for the case of alien oceans. Specifically, let's consider swimming versus running versus flying. In all of these cases, evolution's answer to dealing with predators seems to be go faster and get away. But with running, evolution also eventually turned to tools, and land-bound creatures learned to go on offense: fight instead of flight. Tools were not just tools; they became weapons. The air and sea provide three dimensions of escape, the land just two. Perhaps that is the evolutionary difference between fleeing fast and building weapons.

Again, this is not to say that biology has not explored weaponry as part of Darwinian selection. From squid ink and deer antlers, to bird talons and beetle horns—biology has worked this problem in many different ways.

At some point, however, innovation moved beyond the genetic and into the intellectual. Tools and weapons became something created by the organism instead of by biology. From there, it was off to the races. Once mammals like *Homo habilis* started using non-biological[9] tools and weapons, the path to modern technology was, perhaps, inevitable.

Mobility is also an important consideration for finding food. For example, if you cannot keep pace with the seasons and the changing availability of food, then you need to control food production. You need to farm. Once an organism does not have to flee—once it can defend its territory with weapons and tools—then many options, such as farming, arise.

Consider the alternative. Creatures that can fly and swim may short-circuit the need for growing food. If you can migrate long distances, you can always find new food and breeding grounds. From whales to birds, the ability to migrate may have eliminated the environmental pressure on many organisms to figure out how to farm. If you can cover long distances over the course of a few weeks to months—that is, you can keep pace with the changing seasons—then you can follow the food.

In the ocean and in the skies, covering such long distances is not that big of a problem; you can move in three dimensions and easily avoid obstacles like mountains. You can also move faster—whales and dolphins average 10–20 miles per hour (20–40 km per hour), geese average 40 miles per hour (nearly 70 km per hour), and humans average only about 5 miles per hour (10 km per hour). The most notable land-based exception is the great wildebeest migration in the Serengeti and Masai Mara. However, the animals only cover several hundreds of miles, and they do it largely to follow the rain as the seasons oscillate around the equator.

Agriculture and the need to farm is important because it becomes one of the critical drivers in the development of early technological intelligence. Tools developed for agriculture are tools for stability. If you are stable, then you can store your tools. Your tools become a craft, and the craft of tool-making enhances your ability to farm and stay in one place. The feedback loop begins: with more stability, the tools become better, and as the tools become better, life can become more stable. At this point, you begin to acquire stuff.

Humans are obsessed with stuff. And stuff has served us well. We have made stuff to grow and store food, and we have even made stuff—such as satellites—to examine our entire planet from above. Stuff is at the heart of our civilization.

Making stuff—making tools and developing technology—gets complicated quickly. Fire becomes important. Early humans started using fire a few hundred thousand years ago to as much as a million and a half years ago. Fire was used to cook food and to make tools. Cooking and smoking food were valuable innovations. An ancient ape likely left an antelope leg near a dwindling fire and woke up the next morning to find a tasty cooked meal. That ape's name is lost to history, but I would love to know how quickly the idea of cooked food caught on. It must have been the smartphone of its day: everybody had to have cooked food.

Cooking and smoking meat made it possible to store protein-rich food for long periods and to expend much less energy when digesting meat, as cooked meat is easier to digest than raw meat. With the ability to hunt, cook, and eat more efficiently came "free time." No longer did our ancestors have to hunt and gather every hour of every day. Our brains could grow.

From that free time came new crafts, new experiments, new art, and a lot of innovation. Eventually, we, as humans, arrive at writing and the printing press, the telescope, the radio, the transistor, rockets, and, of course, the Internet. So much *stuff* with so many possibilities.

What I have described above is a highly compressed, and greatly over-simplified, version of how land mammals became intelligent, techno-logically advanced explorers of our solar system. Is there any chance that creatures constrained to an ice-covered ocean might ever become so tech-nologically capable?

MARINE METROPOLIS

In a global, ice-covered ocean, mobility would be in three dimensions; thus if unconstrained mobility removes the pressure to develop tools for fighting and farming, then from the outset the path to technology might be limited. Intelligence may arise, but complex tools may not.

But let's assume, for the moment, that mobility in three dimensions does not preclude farming and developing the tools needed to fight and defend territory. Within these deep, dark oceans, how then, might intel-ligence progress to tools and farming, and ultimately to technology? Again, we are deep in the realm of fun speculation and we are using human civilization as our only available guide.

Perhaps akin to oases of water found in a dry, sandy desert, hydrother-mal vents could serve as deep-ocean oases of chemical energy for pow-ering life along otherwise barren seafloors. The vents, as oases on the seafloor, might provide the foundation for farming.

Much like a river can be rerouted to provide water for fields of crops, the fluids coming from a hydrothermal vent could be rerouted to feed vast swarms of chemosynthetic microbes, which in turn could feed fish, shrimp, and other creatures that farmers in an alien ocean might prefer to eat. Instead of chimney structures, these underwater farmers might modify the shape of the vents so as to control the flow of hot water and nutrients. The vent structures might look like a massive tree with fluid percolating out into bulbs of growing microbes, which could then be harvested like fruit off a tree. Microbial orchards on the seafloor, driven by hydrothermal vents.

Years ago I saw something that made me think I might not be crazy for thinking along these lines. I was at the bottom of the ocean, staring at the chimneys of the Snake Pit hydrothermal vents, 3.6 km down in the Atlantic Ocean. Unlike the other vents that I had seen, which had tall, narrow chimneys, the chimneys at this site were broad, and along the sides were bowl-like structures. These bowls looked like the bottom half of a clam shell and they were made from the minerals of the chimney. They stretched out from the main body of the chimney, giving it the appearance of something that only Dr. Seuss could dream up.

What made this structure astounding, however, was that within these bowls were hundreds, if not thousands, of baby shrimp. Adult shrimp skirted about the chimney, and within each of these bowls were swarms of their progeny. Why were they there? I can only imagine that the vent structure channeled useful compounds from the vents into the bowls to help feed the shrimp[10] and keep them safe. The chimney had been formed into a sort of nursery farm for the shrimp population.

How could that happen? The chimney certainly doesn't care about forming these bowls. Every other chimney I had seen looked like a chimney, not this Dr. Seuss concoction. The shrimp certainly could not build the chimney this way. And yet, through some quirk of biological and geological interaction, this nursery had formed.

Perhaps when the female shrimp around this chimney initially released their hatching eggs, they dropped them into little notches that had randomly occurred on the side of the chimney. The larvae shrimp then had a little shelter, but as they grew, they affected the flow of water and minerals out from the chimney notches. The swimming larvae then caused these rounded bowls to form, as the minerals preferentially grew beneath the mass of swimming baby shrimp. However it occurred, biology—the shrimp—were clearly impacting the shape of this bizarre chimney. If swarms of shrimp can sculpt a hydrothermal vent, perhaps it's not crazy to imagine that creatures could do the same on a distant world.

Unfortunately, even if hydrothermal vents could be harnessed for farming, geology would still impose a limit. If Earth's seafloor is any guide, hot black smokers might not be stable enough for sustained farming. Hot vents often last only a few years to decades before erupting and paving over the landscape with fresh basalt. Along the spreading ridges of our seafloor, heating from Earth's mantle comes and goes over the decades, centuries, and millennia. As one vent system cools down, another may be heating up. When a vent field gets paved over by an eruption, all the life on the vents die. But it only takes a few hardy organisms to restart biology at a new vent site. Microbes are floating everywhere in ocean water, and they start populating new chimneys as soon as the eruption is done. The larger ecosystem is reborn, as tubeworm larvae, muscles, and shrimp settle in on the new real estate. Within a few years, a site can go from eruption to flourishing ecosystem.[11]

If intelligent creatures were ever to develop farming around hot hydrothermal vents, they would need a contingency plan. They would have to pick up shop and find a new site before they get fried by a deep ocean eruption. Like the perils of humans living around Mt. Vesuvius in ancient Italy, it's better to leave early than to miss your chance.

Hot black smokers, however, would not be the only option available for farming. Serpentinizing vents, like those we discussed in chapter 10, are the slow, cooler, and more stable sister vents to the black smokers. The exothermic chemistry that powers these vents can trickle along for many thousands of years. The Lost City hydrothermal vent site, for example, is at least 30,000 years old,[12] which is about 8,000 years longer than we humans have been farming. Perhaps there is some hope for the early stages of farming and civilization within the depths of a dark ocean.

If hydrothermal vents like Lost City or even Snake Pit exist in alien oceans, might some alien octopi, in our solar system or beyond, have realized the farming potential of deep-sea vents? Might they be growing vast cultures of microbes to feed the food chain of their civilization? I like to think so. There might also be similar but inverted systems at the base of the ice shell. Brinicle drainage points from the ice could be plumbed to channel the rich oxidant and sulfur chemistry into icy farms for feeding fish. These two systems of channeling chemical energy—one type at the seafloor and another type at the base of the ice shell—could be the ocean world analog of agriculture and farming.

Beyond farming, another consideration is cooked food. Earlier in this chapter I mentioned the importance of cooked food for early humans: cooked meat is easier to digest, freeing up energy that can be rerouted from the digestive system into the brain and nervous system. In addition, cooked and smoked meats last longer than uncooked meat, providing a more stable source of food to get through cold winters and long droughts.

While an open fire is clearly not possible in an ocean, it is interesting to consider whether or not cooked food is possible. The high temperatures of hydrothermal vents, and possibly the hot rocks in fresh lava flows within the ocean, could be used for cooking food. Dangle a fish over the mouth of a vent and you could boil it in the super-heated water. To the best of our knowledge, no one has ever seen any "cooking" behavior around hydrothermal vents on Earth, but that's not to say that cooked food wouldn't have a dietary advantage.

The biggest challenge for sustained cooking would be the lack of control. Early humans eventually figured out how to start fires of their own, instead of capitalizing on the occasional lightning strike or forest fire. In the best case scenario, deep-ocean creatures would have a reliable hot vent, but I cannot see any way they could initiate a cooking mechanism of their own, akin to starting a campfire. Roughly half a million years ago, a clever cave man or woman, figured out how to rub sticks together, or bang stones together, to get the spark needed to start a fire. From that innovation came the stability of fire. Even the most brilliant octopus in the galaxy will never have the good fortune of seeing fire start in that way. The heat needed to cook food could be a limiting factor for the development of big brains in deep oceans.

And that's just the beginning of the challenges. While the senses and sensory perception could lead to intelligence, that intelligence may have a hard time making the leap to technology. Farming and cooking are important, but they are still a long way from a technologically advanced civilization.

As humans developed increasingly capable tools, we graduated from the Stone Age (which ended a few thousand years before the common era [BCE]), to the Bronze Age (which lasted, in some regions, until a few hundred years BCE), and then into the Iron Age (which extended from the end of the Bronze Age into the first few hundred of the Common Era, again with regional variations).

The advent of metalworking was a central component of our ability to innovate more complex tools. An agrarian society in the deep ocean would have access to stones and the tools that could come from modifying stones. If anything, the fresh lava flows could provide easy access to sharp, glassy stones that could be used as early knives and cutting tools. Sure, working in the viscous water environment of the ocean makes swinging an axe or hammer more difficult, but there are numerous examples in marine biology of hammering or whipping motions used for hunting (e.g., shark tails and squid beaks). Working with stone tools, and having a "Stone Age" in the deep ocean seems possible, in theory.

But metalwork? How could a civilization move into a Bronze or Iron Age without fire and furnaces to smelt different metals together? What would it mean to be a metalsmith at the bottom of the ocean? This could be a severe bottleneck for the advancement of technologies in ocean worlds. Cleary, I have only our human example from which to draw, and thus I may not be able to conceive of other solutions. With that bias in mind, let me try to imagine a way forward.

For humans, the advent of the Bronze Age may have been contingent on serendipitously mixing copper- and tin-rich stones in a fire pit (bronze is an alloy of copper and tin). Once heated to about 950 °C, those stones might have oozed out bronze. The lucky humans who found this strong but malleable metal when the fire went out would have been delighted with its utility.

In the deep ocean such metallurgic serendipity might not be impossible. The hot, black smoker hydrothermal vents are full of heavy metals, and they are also near an intense heat source. In many cases, the mineral mines that you see on land were once hydrothermally active regions in the ocean. The chimneys and large deposits that flow out of the vents are the precursors to many of the veins of precious metals we dig so deep to mine.

Could a curious alien octopus accidentally drop a few copper- and tin-rich stones near a lava flow, causing them to smelt into bronze at the bottom of the ocean? It seems unlikely but so does the idea of some hairy mammal building an airplane.

Perhaps the initial entry into metalwork at the bottom of the ocean could be driven by the richness of hydrothermal deposits and the heat of lava flows. But to establish a sustained industry of metalwork would require better control over the heat source. No smart octopus wants to be constantly risking their life near a lava flow. Is there any way to have a controlled heat source underwater?

On the one hand, welding torches and high temperature chemical reactions can work underwater. On Earth there's a healthy industry dedicated to welding ship hulls and fixing structures under water, largely using arc welders. Arc welding uses electricity to create the high temperatures needed to melt metals.

On the other hand, I have solved one challenge by creating another. We humans started working with metals long before we discovered electricity. If metallurgy and metalsmithing in the ocean requires electricity as a heat source, it's probably game over for technological advancement made possible through metallurgy.

Jumping even further forward in technological evolution, we can ask if electricity is essential in the same progression we humans experienced. Does a civilization need bronze and steel to eventually get to electricity, and ultimately computation? What would it take to get to that most essential of electronic components: the transistor? What would a transistor look like if it were developed by an advanced civilization deep in an alien ocean? For that matter, is computation possible in the ocean?

These, and many other questions are fun to ponder when thinking about evolution on ice-covered ocean worlds. There may be strong limitations to how far technology can proceed in deep oceans, but I think that biology may provide some of the solutions. Our human bias, and our technological path, might be leading us astray. Perhaps there is another way.

TECHNOMIMICRY

If the conditions are right, I think evolution could proceed to the point of intelligence and possibly even some form of farming on the seafloor, and along the ice-water interface, of alien oceans. But when it comes to more advanced technologies, and the development of those technologies in a progression similar to our own, I do not see an obvious path. Fire and the controlled manipulation of metals, ceramics, plastics, and other materials is simply not possible when you are immersed in a liquid water environment.

However, when we think about technology, we think of it as something other than "natural" or biological. That makes sense—we have created much of what we see in our civilization with our brains and our hands.

But what if some of the innovations we consider as "technology" could evolve biologically? We humans might be a bit of an anomaly in that we, to use an appropriate pun, short-circuited biology and jumped to technology by harnessing fire and electricity in our external environment. What if electricity and computation in biological systems could develop into much greater complexity than we have seen in life on Earth?

In modern engineering, when a design takes inspiration, or even mimics the details of a biological solution, it is often referred to as "biomimicry."[13] Many of our brightest and most innovative engineers look to biology for inspiration. Mollusks grow intricate shells in cold ocean water with nothing but the dissolved gases in the water and the stuff that drifts on by. Trees grow thousands of leaves, each of which is more efficient at harnessing the Sun's energy than our mass-produced solar panels. For us to make ceramic materials and solar panels, we have to undertake complex industrial processes in factories; yet biology does it all the time, all over the planet, under many different conditions. Biology has figured out some very elegant solutions to many challenges.

I like to think that the reverse of biomimicry might also be true: that many types of technologies we have developed have analogs in biological systems somewhere in the universe. I like to think of this as "technomimicry": biology mimicking something we only discovered through technological innovation. It may be that we are underestimating Mother Nature when it comes to the line between biology and technology.

Consider computers. Computation and the ability to store information for long periods of time with very high fidelity has obvious survival value. Computers make it possible to track changes in the climate of your planet, and make predictions about the future. Anything that helps an organism predict the future has survival value. Computation, I think, could also lend itself to technomimicry.

The human brain is capable of complex calculations, and it can quickly process data. But our memories fade and can be low fidelity. We can drive cars at 100 mph, but we can't predict the weather a week from now. Computers, meanwhile, have proved useful for many of the tasks our brains cannot handle. Over the past 50 years, we have used our ability to innovate technologically to compensate for the shortcomings of our biological brain. It's a process that started with cave paintings and has brought us the Internet.

As a specific example of possible "technomimicry," consider an external hard drive. Could biology, through natural selection, arrive at the equivalent kind of data storage used on a flash drive? A flash drive, or Universal Serial Bus (USB) drive, uses flash memory to store information, even after the power is turned off or the flash drive is unplugged. This kind of memory is also called "non-volatile" memory, meaning that it won't vanish after power is lost. You may have heard the term EPROM, or erasable-programmable read-only memory, which is another name for this kind of capability.

Obviously, our brains have some degree of non-volatile memory. Our cortex can store memories for a lifetime, and when we go to sleep, we do not lose everything that ever happened to us. But even in sleep, our brains use a little power. Much like a computer, a biological organism could have an evolutionary advantage if at least part of its brain was capable of storing data without needing power. In theory, it might be possible for evolution to arrive at an analogous system.

Within a flash drive are vast arrays of tiny transistors. Transistors form the basis for nearly everything in computing; they are like the atomic unit of computation, and they can be combined in different ways to make a variety of things possible. For the sake of understanding non-volatile memory, all you need to know is that transistors are made of semiconductors, and semiconductors are simply layers of silicon doped with small quantities of different elements to fine-tune their ability to conduct electricity.

The simplest version of a transistor works like a drawbridge. In this analogy a small, low-powered car is used to pull on a rope that lowers the drawbridge, which starts the flow of traffic across the bridge. The drawbridge is spring-loaded to keep it in the up position, and thus if there is nothing pulling on the rope—no tension on the rope—then the bridge is up and the flow of traffic is stopped.

Electrically, the small car is like the small current in a transistor that is used to start the flow of a large current through a sandwich of three semiconductors. Information can be stored in the form of a 0 or a 1, which corresponds to the drawbridge being up or down, respectively.

When power to this simple form of a transistor is lost, it's as though the rope to the small car is cut and the drawbridge automatically moves to the raised position. No traffic can flow, no current can flow, and no information can be stored because all of the transistors read 0 in this situation.

To make a transistor that *can* store information without power, electrical engineers added a little piece of semiconductor material to the middle layer of the basic three-layer transistor sandwich. That extra little semiconductor is electrically isolated when the power turns off, and thus it can retain the charge of any current that was running through it at the time it was powered off. It's like having a second small car attached to another rope on the drawbridge, and the tension on that rope is connected to a ratchet that retains the last state of the drawbridge before the power is lost. Even though the drawbridge goes up when the power is cut, the tension on the second rope is saved; and when the hard drive is powered up again, the drawbridge moves up or down depending on the tension in the rope. This is how data—0's and 1's—are saved even when the power is off.

This innovation on the basic transistor is called a floating-gate transistor, which is a specialized type of metal–oxide–semiconductor field-effect transistor (that is, a transistor with a metal gate, oxide insulation, and a semiconductor). The "floating-gate" refers to that extra piece of semiconductor that can store the charge state (i.e., the tension in the rope) without power for long periods of time.

For biology to have an adaptation that leads to flash memory, there would likely need to be some semiconductor chemistry embedded within the nervous system. Specific types of neurons might use extremely small and carefully grown grains of doped silicon to create the functionality of floating-gate transistors. Perhaps such neurons would have their origin in microbes, which would later symbiotically incorporate into the nervous system. Many interesting examples exist of complex minerals being made by microbes in their environment.[14] On a world like Europa, such microbes might initially be working to extract energy from the changing electric and magnetic fields in the ocean. Like the microbes on Earth that later merged with larger cells to become mitochondria, chloroplasts, and the ribosome, these electromagnetically skilled microbes might find their talents acquired by a more complex biological system.

The incorporation of biological transistors and a form of flash memory into a nervous system would improve an organism's ability to store information about its environment and compute predictions about the future state of its environment. This is potentially of great survival value over the long-term evolution of an intelligent organism. Again, this is all highly speculative but intriguing to consider.

To conclude, when it comes to the octopus and the hammer, advanced technologies that would require metallurgy may not be possible, but the path of human civilization and technological development might not be the only way. Technomimicry could provide a different path for arriving at similar capabilities.

Perhaps if intelligent and technologically advanced organisms exist, they would grow large, shimmering abalone shells that form domes over hydrothermal vents, capturing the chemicals that come out and controlling the flow of energy from the seafloor to the surrounding ocean. Across the thickness of the shell would be a flow of electrons and protons, and from that flow they might establish a stable source of electricity, and perhaps even bioluminescence.

Within these biodomes would be vent farms that feed much of the community. But the need for oxygen from the ice shell above would require a conveyor belt of sorts. From the domes and up to the ice would be a parade of creatures that use the vent fluid to power floating bladders that rise up through the ocean and deliver them to the ice shell, where they then mine high-concentrations of oxygen and sulfate for use on the farm below.

And within that ice shell—returning to the Gaia concept—would be microbes that regulate the size of the ice grains and regulate the rate at which the ice brings radiolytically produced nutrients from the ice surface down into the ocean below.

Biology would control the cycling of the ice shell, and maybe even the location of the hot spots on the seafloor, because establishing stable sites for the domes would be highly advantageous. Organisms whose job is to grow extremely deep roots into the rocks might enhance the longevity of a serpentinizing hydrothermal vent by permitting more water to spread deeper into the rocks, where the chemical reaction could proceed with fresh material. Perhaps even advanced computation exists in and among the nervous systems of such organisms, allowing them to develop models for predicting the behavior of their tidally stretched moon.

And yet, despite the potential for a technologically advanced species, creatures on such a world would have no night sky to foster their sense of wonder. They might never be able to sense beyond the ice to discover the stars above. Could they ever develop the impulse to explore beyond their own planet, not having the Sun or stars to compel them into the sky?

Think about what that might mean for their philosophy, their art, their music, and their concept of the meaning of life.

For centuries our gods lived among the stars. Religion and mythology on Earth have deep roots in our connection to the night sky—from the star of Bethlehem to the daily horoscope, the night sky has long been a backdrop for our culture and civilization. Heaven is above, hell is below.

Imagine being an intelligent creature within the ocean of Europa, Enceladus, or Titan, or even some ice-covered ocean elsewhere in the galaxy. The ice is too thick to allow any light to reach the ocean. No stars beckon your curiosity and motivate your mythology. Perhaps the ice creaks and cracks with a certain rhythm. Sounds from this creaking and cracking chorus form the basis of your mythology. The myths would be connected to acoustics—i.e., sound from the cracking ice shell—instead of to light from stars above. On a world like Europa, the tidally flexed and cracking ice shell would send sounds throughout parts of the ocean, and the creatures within—if they were intelligent—would be perplexed by these sounds and the regularity of that pattern. Little would they know that Europa's sounds were connected to the massive planet Jupiter, around which Europa is tugged as it orbits every 3.55 days.

And little would they know that Jupiter orbits a glowing orb of burning hot plasma once every 12 years. Nor would they know that that orb, our Sun, is only one star of many billions in our galaxy and that our galaxy is one of many billions in our universe. They wouldn't know that nestled next to that orb is a tiny rock with a dab of water covering much of its surface. They would not know that curious creatures evolved on that rock—creatures with the ability to imagine and innovate, creatures with minds that would compel them to explore every frontier, creatures with a long-standing desire to know whether or not they are alone in the universe.

In my late-night gazing at the sky above, I like to imagine that within some of these alien oceans beyond Earth there are in fact biologically, and perhaps technologically, advanced civilizations. Given the abundance of ocean worlds in our solar system, such worlds might be ubiquitous throughout the universe. I only hope that if they are out there, if there are intelligent creatures deep beneath the icy shells of these distant worlds, that some of them have found a way to peer through the ice and see the wonder of what lies beyond.

CHAPTER 13

A PERIODIC TABLE FOR LIFE

Over the last several chapters we explored the science of how life works and what it might take for life to arise on ocean worlds beyond Earth. I have not, however, made much mention of the actual discovery of life, and how it might change the way we think about biology and our place in the universe. In other words, I have yet to address the "So what?" of such a discovery. This is a question I am often asked, and I think it is incredibly important, especially when much time and money is required to explore our solar system.

To begin with, the discovery of life beyond Earth—or, alternatively, the revelation that we are alone in the universe—will have no obvious short-term impacts. It's not going to change the way you make your coffee in the morning or shorten your commute to work. It won't provide a cure for cancer or stop climate change.

But it will revolutionize our understanding of biology. Biology is the science of life; it is the phenomenon of us, and we do not yet know whether it is a universal phenomenon. As I've mentioned elsewhere in this book, we have yet to understand whether or not biology works beyond Earth. Other major scientific fields have benefited from looking beyond Earth to learn more about the scope and underlying principles of the field. Looking to the stars and planets initiated the Copernican Revolution and changed forever the field of physics. The field of chemistry was, in part, catalyzed by studying light from the Sun, as we covered in chapter 3. Geology, and the principles of geology, were transformed by the detailed study of meteorites, rocks returned from Moon missions, and our robotic exploration of Mercury, Venus, the Moon, Mars, and asteroids. Biology has yet to make that leap. I suspect that the science of biology will also experience a similar expansion of understanding through exploration.

Chemistry can perhaps be a useful guide for considering how much we have yet to learn in biology. Decades before Dmitri Mendeleev developed the first version of the periodic table, elements such as oxygen had been isolated and characterized, but the true meaning of an "element" was still a vague concept. As more elements were discovered, the relationships gradually appeared. Observed behaviors could be related to quantities like atomic number and mass. Organizing the elements by similarities in behavior, such as affinity for reacting with other elements, gave rise to the rows and columns of the periodic table. From that emerged an understanding of the importance of the electron valence—the "fullness" of the outer electron shell—and how elements from one column preferentially bond with elements of another column.

We have come a long way since those early days of characterizing the elements, and we now have a complete framework for how the elements relate to one another. The map of the elements—the periodic table—also serves as a recipe of sorts for the science of chemistry.

I wonder if we might someday have a periodic table for life; a framework for relating and connecting the many different forms of life that we may discover in our universe. Life on Earth—water, carbon, DNA, RNA, ATP, and protein-based life—would be only one piece on the galactic quilt of biology's fundamental map. From that quilt we would begin to see the contingent and convergent qualities of biochemical evolution and life under a range of planetary conditions. We might begin to understand the fundamental principles of life and perhaps answer that timeless question, "What is life?"

The periodic table of elements, however, was born from a reductionist approach: keep breaking things apart until the smallest pieces are revealed. The atomic theory of matter—a prerequisite for developing the periodic table—made that possible.

The reductionist approach for biology does not work. If you take life and try to split it apart until you have the "basic" components, you destroy that which you are attempting to study. The smallest "unit" of life is the microbial cell. If you take a cell and break it into pieces and then reconstruct it as it was, you will end up with something that looks like a cell, but it will be a dead cell. The "life" will be gone.

This is not to say that there is something magical about "life," but rather that life is more of a process than a tangible "thing." As microbiologist Lynn Margulis once said, life may be a noun, but it's really also a verb. Biology could simply be a collection of matter that is "life-ing."

To help frame our thinking about the possible diversity of life in the universe, we step through several key parameters for life as we know it. These are parameters required for life on Earth, but they could also be varied to include a host of chemistries for "weird life."

SUITABLE SOLVENTS

At the highest level, life needs a liquid. It needs a solvent in which to dissolve compounds and host reactions. Our cells, and every living cell on Earth, is primarily water, which is the solvent for all life on Earth. The dominant phase of matter for all life on Earth is liquid water.

There is perhaps the possibility that life forms out there could be made of predominantly gases, solids, or even plasmas (the stuff of stars). I cannot rule it out. But the keystones of life that we have covered throughout this book heavily favor life that is based on some form of liquid. In solids, it is too hard for molecules to move around, and thus reactions are inhibited. In gases, the molecules move around too much, and they may not bump into each other often enough for reliable reactions to occur. In liquids, it seems, chemistry achieves a balance that is particularly conducive to the needs of life. Life needs a solvent.

With that in mind, the next consideration is whether or not there are liquids other than water that could be suitable for life. Plenty of liquids exist throughout our solar system. On Earth alone we've got acids, such as sulfuric acid and hydrochloric acid; we've got various forms of alcohol; we've got ammonia; and we've got organic compounds, large and small, that can form liquids. Next time you're in the cleaning section of the supermarket, look at the variety of chemicals available on the shelf. It's a fair representation of the options biology might have. Sulfuric acid is plentiful on Venus and perhaps even within the cracks of Europa's ice shell and the ocean itself. Ammonia, meanwhile, rains out of the clouds of Jupiter and is found on the icy surfaces of Triton, Pluto, and likely many other Kuiper belt objects.

One way to classify liquid solvents is by the polarity of their constituent molecules. We covered some of this in chapter 7 when we considered the case of "weird life" in the liquid methane and ethane lakes of Titan. Water molecules have a slight negative charge on the oxygen end of the molecule, and a slight positive charge on the hydrogen ends. In other words, each water molecule has a positive and negative pole, and thus liquid water is polar. These poles result from the oxygen atom "borrowing" the single electron from each hydrogen atom in order to fill its outer shell of electrons. If electrons are shared equally across all atoms in a molecule, then that molecule is non-polar.

The polarity of water makes it a great fluid for dissolving other polar molecules, such as many of the amino acids that are important for building proteins. As a solvent for life, water helps with the task of assembling polar molecules and pulling them apart when needed. The carbon-based life that we know and love works entirely on the principle of polar carbon molecules moving, mixing, and combining in the polar solvent of water. Amino acids linking together to form proteins is only one important example.

Methane and ethane molecules, as we learned, do not have polar ends, and thus the liquid form of these compounds is non-polar. Like dissolves like, and thus non-polar liquids are good for dissolving non-polar compounds. A few examples are the compounds acetylene and ethylene, both of which may be dissolved in the lakes on Titan. Most hydrocarbons, which are made solely of hydrogen and carbon—no oxygen, nitrogen, sulfur, etc.—are also non-polar.

An easy way to think about non-polar compounds is to consider liquids that do not dissolve in water. Oil and water don't mix; all the compounds in oil float instead of dissolving. Many of those molecules are various forms of saturated and unsaturated fats. Our cells, and the cells of all life as we know it, are fancy bags of salty water. The bag itself—the membrane—is made, in part, of lipid molecules with a non-polar, hydrophobic end that keep the cells together. It could be that many forms of life, perhaps on Titan and other worlds, function with cells that are fancy bags of non-polar liquids, like methane and ethane, encased in a membrane of polar molecules.

Beyond the polar versus non-polar split for solvents, let's consider additional attributes of any solvent for life. One helpful characteristic of water molecules that is largely derived from its polarity is the affinity that water molecules have for bonding with other water molecules. We see this every day in the form of surface tension in a glass of water, or raindrops on a windshield.

Not every polar solvent has this same quality. Ammonia, for example, could turn out to be a great solvent for life, but based on our earthly assessment, it falls short. The ammonia molecule has three hydrogen bond donors and only one acceptor; water has two of each.[1] As a result, water molecules link together like a microscopic chain-link fence, but ammonia, although polar, is not as good at separating from polar organic compounds like oils. This is bad for compartmentalization and for making effective membranes. Ammonia, which is a basic solvent[2] with a high pH, can grab a proton off an oil molecule, which gives that molecule a negative charge. With that new charge, water molecules will now bond to the oil molecule. This is great for cleaning surfaces, as ammonia and water working together can pick up oil spills, but ammonia on its own is perhaps too good at reacting with other compounds.

Nevertheless, ammonia is abundant in the outer reaches of the solar system. As we explored in chapter 8, Triton and Pluto may have subsurface oceans rich with ammonia. The chemistry of weird, ammonia-based life could be chugging away on these worlds without any care for my earthly bias and failure of imagination.

Lastly, I note that one cannot strictly rule out other curious but very abundant liquids, such as liquid hydrogen, which is found deep within gas giants like Jupiter. Dr. Steven Benner and colleagues examined this prospect and concluded that, if the temperatures are not too hot (not over

500 K), there is a chance that some interesting carbon chemistry could occur within liquid hydrogen.[3] Whether or not that leads to life within these worlds is a question, unfortunately, that will remain out of the reach of robotic exploration for quite some time.

The relationship between the element carbon and various solvents is very interesting. The incredible array of compounds that carbon can make leads to some that are polar and some that are non-polar. Carbon can make compounds that dissolve within the water of the cell while simultaneously building the molecules of the membranes that hold everything together.

This is part of why carbon is such a useful element for building life—which brings us to the next big consideration for the biochemical diversity of life in the universe.

THE BEST BUILDING BLOCK

Beyond the solvent of life, the next most important parameter for classifying biochemistries is the basic building block, or central element needed to build life. For all life on Earth, that element is carbon.

Carbon is hands down the best team player on the periodic table. It has four electrons in its outer electron shell, but it wants a total of eight to fill the shell. This means that carbon is willing to share its four outer electrons as long as it gets to simultaneously borrow four electrons from other elements.

As an example, when forming methane (CH_4), carbon teams up with four hydrogen atoms, each of which gets partial ownership over one electron from carbon in exchange for sharing its single electron. When forming carbon dioxide (CO_2), however, each oxygen atom gets allocated two electrons from carbon in exchange for sharing two of their own electrons.

Carbon cooperates and forms bonds with many elements across the periodic table. The covalent bonds that carbon can form with a wide variety of elements make it ideally suited for the construction of large, complicated molecules that are stable but not impossible to break. This is important. In order to replicate and reconstruct itself, life needs molecules that are stable but not too stable. The bonds that form should be strong enough to hold life together, but not so strong that they cannot be split during the processes of replication and reproduction.

When looking at the periodic table, it is reasonable to expect that by moving one position down from carbon you should arrive at an element of comparable character. You should, the table tells us, arrive at an element that carries a similar affinity for forming bonds with other elements. Below carbon, lies silicon. If carbon can serve as the backbone of life, so should silicon. Judging by its position on the periodic table, one might rightfully assume that silicon-based life could follow in the biochemical footsteps of carbon.

Silicon does indeed form some very good bonds with other elements. So good, in fact, that those bonds form much of the ground on which you stand. When silicon bonds with oxygen, it usually forms silicate, which when combined with iron, magnesium, and any number of other elements forms the rocks beneath or feet. Silicate rocks are the stuff of which rocky planets are made. The chemistry of silicon—the chemistry of minerals—is largely the chemistry of planetary geology. The crystals of many minerals are driven by the tetrahedron shapes of silicon bonded with oxygen. Our experience so far with silicon and carbon chemistry is that *geology* leads to the strong, solid structures of silicate mineralogy, while *biology* leads to the folding carbon structures of deformable proteins.

This is not to say that silicon does not form structures that are compelling from a life standpoint. Silicon crystals in rocks are capable of forming long chains and sheets of connected elements, but the temperature and pressure at which those silicon-oxygen bonds break is too high for them to be of much use for life. Crystals are either strong solids or molten hot liquids.

Although silicon may have some limitations when it comes to forming silicon-based life, it is useful to consider minerals, and the way they are classified, as a possible framework for classifying life in the universe. Minerals are defined as inorganic, naturally occurring solids with a highly ordered atomic arrangements and definite (but not fixed) chemical compositions.[4] It's not a particularly concise or elegant definition, but it gets the job done, for the most part. Many minerals have properties, such as hardness or luster, that only occur on the larger scale of a complete crystal. Break a mineral apart into its elements, and it no longer has those intrinsic properties.

Roughly 3,800 mineral species exist, which is much greater than the 118 elements of the periodic table. Mineralogists have organized the minerals into 12 classes, and those are divided into families, groups, and species based on the chemistry and physical properties. Within species there can be an even further division into varieties. Silicates comprise the largest class of minerals by far. Silicon in the mineral world is comparable to carbon in the biological world. Perhaps someday we will find that the diversity of biochemistries used by life in the universe follows a classification pattern similar to that of geochemistry and minerals.

Returning to the topic of the best building block, another important consideration for the key element of life is the conversion into different phases. As an element is consumed and then exhaled or excreted, a phase change helps to move it in or out of an organism. Carbon dioxide is a great example. Many microbes make a living by consuming carbon dioxide, which is a gas (in the atmosphere or dissolved in water), extracting the carbon and using it to build solid structures in the cell. Plants also do this during photosynthesis.

Meanwhile, other microbes and organisms (we humans included), consume carbon-rich materials and convert it into carbon dioxide, efficiently getting rid of the carbon dioxide "waste" as a gas. When it comes to life, the various molecules that carbon can form include solids, gases, and compounds that dissolve well in liquid water. This versatility is very useful for life, particularly for organisms that cannot move (e.g., microbes) and which depend on the environment to bring in new food and to get rid of waste.

Silicon does not fare very well by this standard. While it forms very stable solids and forms liquids at high temperatures and moderate pressures (as evidenced by lava flow from volcanoes, which are predominantly molten rocks made of silicon, iron, and magnesium), there is no real gas phase version of silicon. For example, silicon dioxide (SiO_2), which would be the analog to CO_2, is a solid. In fact, it's a solid with which you are very familiar: it's glass. At best, silicon dioxide's close relative, silicic acid can exist in dissolved form in liquid water. On Earth, the dissolved silica in our ocean makes it possible for microscopic diatoms to build intricate shells, but that is the extent of the use of silicon.

Over a wide range of temperature and pressure conditions relevant to planets and moons, silicon is either locked up in solid minerals or it is moving and flowing in a very hot magma. I cannot say for certain that life forms do not exist in liquid magmas, but so far nothing has crawled out of any lavas on Earth, and I seriously doubt anything could.

Part of carbon's versatility is because a variety of carbon–carbon bonds are possible. Silicon typically likes to bond to oxygen, but it can bond to itself, forming long, chain-like structures called polysilanes.[5] These molecules are the closest that silicon comes to emulating the long, diverse chains that carbon can make and that are so important to life. Polysilanes are typically formed in the lab, but it's not unreasonable to think that they could form in nature. One limitation, however, is that the silicon-silicon bond is about 30% weaker than the corresponding carbon-carbon bond, and thus the polysilanes are less stable and quicker to break apart.

Interestingly, it has been argued that silicon chemistry could be preferable in very cold liquid environments,[6] particularly in liquid nitrogen (-320 °F), which could exist within the interiors of worlds like Pluto and cold, distant Kuiper belt objects. Along with having a low-bond energy, silicon is more reactive than carbon, in part due to the arrangement of electrons in its outer shell. Thus it could potentially be a better element to use when building structures at low temperatures, where reactions are already slowed down considerably.

Beyond silicon, the periodic table has only a few other elements in the same group as carbon, and none of them have the versatility of carbon. Germanium, tin, and lead are all fine elements in their own right, but they cannot bond as well with themselves as carbon does. As is the case with silicon, this is a limiting factor for forming the polymers of life. Even under a range of temperature and pressure conditions, these other elements simply are not the stable team player that carbon is.

In a twist to all this chemistry, in recent years scientists and engineers have become much more adept at using biology itself to engineer new compounds and new solutions to challenging problems. Nobel laureate Frances Arnold, at the California Institute of Technology, runs a premier lab on this topic. Her team managed to get enzymes to do what Mother Nature had not yet done in an organism: form bonds between carbon and silicon. Using microbes and directed evolution, Arnold and her team found that they could make these bonds 15 times more efficiently than standard industrial processes. Their work has obvious application to manufacturing and a host of industries, but I wonder if they might have opened a small window into a type of coordinated carbon-silicon chemistry that could evolve on some distant world. If life on Earth can be taught to use those bonds, perhaps life elsewhere has already figured it out.

THE ARCHITECT AND THE FOREMAN

Moving beyond the solvent and the core essential element, the next level where we might see some diversity in life is the molecular machinery for how information is stored and retrieved.

For all life on Earth, that functionality is provided by DNA and RNA. Earlier in this book I made the comparison of our biochemistry to a construction site for a building: the DNA in every cell is like the architectural blueprint, while RNA functions like the foreman on the job, making sure construction goes well.

The DNA molecule stores genetic information by using four different molecules that help build the twisted ladder of DNA's structure. Those molecules are the nucleotides guanine, cytosine, adenine, and thymine. Like the 0's and 1's of data in a computer, the pattern of these nucleotides in DNA determines everything that is needed for life. From microbe to mammal, it's all encoded with the same system.

Meanwhile RNA has the job of reading, transcribing, and translating the DNA into a form that can be used to build the structures of life—the proteins, membranes, and other machinery. The foreman looks at the blueprint, reads it, and figures out how to create the house that is described. The foreman does this with a team of workers—carpenters, masons, electricians, plumbers, etc.—who each takes a subset of the main blueprint and produces their part. The team is like the enzymes that life uses, each taking a subset of the main DNA (the main blueprint) and making proteins from those instructions.

The DNA-RNA paradigm works well on planet Earth, and we do not know if viable alternatives exist. On the early Earth several different biochemical alternatives might have arisen, but over time they were outcompeted by the success of the "RNA world" (covered in chapter 10). Nature has, perhaps, been running this experiment many times in our own solar system, and the ocean worlds could help us understand the relationship between contingent and convergent drivers toward the biochemistry of DNA and RNA.

Although conceiving of alternatives to DNA and RNA is challenging, framing these molecules in term of information theory is useful. Computers store and manipulate information based on the binary system of bits. Everything you've ever done with a computer traces back to 0's and 1's stored in transistors.

Likewise, DNA does everything by using its four different nucleotides; instead of 0 and 1, it codes with guanine (G), cytosine (C), adenine (A), and thymine (T). From a computing standpoint, DNA has four coding options instead of only two; in other words, while computers use a binary (base-2) system, biology uses a quaternary (base-4) system.

From a simplicity standpoint it might seem odd that evolution and natural selection would lead to a quaternary system instead of binary. With only two base pairs, DNA could encode all the same information as the four-base system and it could avoid the added complexity of those extra two bases. But the downside is the size of the molecule needed to encode that information. Sure, the molecule is simpler, but now it has to be a lot longer, since it cannot store as much information per bond. How much longer? The total amount of information that can be stored is roughly equivalent to the number of different variations possible. For a computer chip with eight transistors, 2^8 or 256 variations are possible. Equivalently, a short strand of DNA with a string of eight base pairs leads to 4^8 or 65,536 variations. To put that into practical terms, in the first case I could encode up to 256 different colors in those eight bits, whereas in the second case I could encode for over 65,000 different colors.

As another real example, consider the number of amino acids that we use to build proteins. The information to build amino acids is coded into our DNA. Life uses 22 different amino acids, which means that if we had a DNA binary system, five molecules would be required to code for one amino acid. A sequence of four molecules would result in 2^4, or 16 options, which is less than 22. A sequence of five molecules is 2^5, or 32 options, which is enough to code for all 22 amino acids. However, with our quaternary DNA system, it takes only three nucleotide molecules to store the information needed for any amino acid. This is because 4^3 is 64, giving DNA all the storage and redundancy it needs.

While the quaternary nucleotide system is more complex, a lot more information can be stored in a smaller molecule. This is highly advantageous for reading and replicating DNA: the longer the molecule, the more likely mistakes are going to be made, and mistakes are very costly in evolution. Ultimately, natural selection works to optimize the fidelity and fitness of an organism. The added energetic cost of working with four base pairs is offset by DNA molecules being smaller and replicated with greater efficiency.

It is possible that billions of years ago, life in the age before the RNA-world might have used a binary system. That binary information molecule could have been the precursor to the more complex quaternary system that we now have.

It is also possible that the quaternary system we have today is not the end of the story. Life on other worlds might employ an octal system with eight different molecules, or a decimal system with ten different molecules. The possibilities are limitless, as long as the evolutionary cost of maintaining a more complex system is offset by the survival advantage.

Recently, a brilliant team led by my colleague Steve Benner, at the Foundation for Applied Molecular Evolution, demonstrated that they could construct an octal DNA system.[7] Along with the standard four bases, they added four more. I'll spare you the long chemical names of the new bases; the team referred to them simply as P, Z, B, and S. In this new form of DNA, the rungs of the double helix are made by linkages between pairs: G and C, A and T, P and Z, and B and S. It was quite an extraordinary development. The information density of the octal DNA dwarfs that of our standard quaternary DNA. Instead of 65,536 variations in an eight-strand DNA string, it could store any one of 16,777,216 options. Benner and team anointed their creation "hachimoji DNA." *Hachi* means eight in Japanese, and *moji* means letters.

While hachimoji DNA is still in its early days and we don't know yet what this discovery might bring, it is a powerful proof of concept that alternatives are possible to our standard DNA and RNA paradigm. On Earth, the temperature, pressure, and other environmental conditions have clearly favored the system we have. But colder environments provide increased stability for large molecules, and thus it is possible that a simple binary coding system with larger molecules could persist within ice-covered ocean worlds. High pressures, such as those found deep within Europa's ocean, likely decrease the chemical stability of a molecule like DNA.[8]

Whether or not conditions within ocean worlds could lead to a higher base system like hachimoji is impossible to say, but evolution will always select for the robust but parsimonious option. Biology's architect and foreman will always need to be the best-suited molecules for the job; Darwin wouldn't have it any other way. In different planetary environments the combination of robust and parsimonious could lead to a variety of information-coding strategies.

BRICKS AND MORTAR

An architect and a foreman can get little done without a good construction crew and building materials. Similarly, DNA and RNA have the information and the know-how to build life, but they need the bricks and mortar to actually put it all together.

Amino acids, nucleic acids, sugars, and fatty acids are the basic bricks that all life on Earth needs. The mortar is the bonds that form when you link these small molecules together. Chains of amino acids form proteins, chains of nucleic acids form DNA and RNA, chains of sugars form polysaccharides,[9] and chains of fatty acids form the lipids in our cells and membranes (with the exception of archaeal microbes, which use the molecule phytane). These small molecules are the monomers that can be linked together to form the polymers of life. The root "mer" comes from the Greek *meros*, meaning "part." Life is systematically built from a select group of small "parts" that enable the construction of complex and specific large molecules.

Life on any world will almost certainly function by taking a relatively small set of specific monomers and combining them into the larger polymers needed by the organism. Proteins serve as a good example for life on Earth. Proteins comprise a diverse and capable suite of molecules, and yet they are formed from a relatively small set of simple molecules. Hundreds of amino acids exist in nature, but life as we know it uses only 22. From these 22 amino acids, proteins are made that range in size from a few dozen amino acids to tens of thousands of amino acids.

The construction of proteins from amino acids is a very powerful chemical code that yields all the functionality life needs, from structures in the cell to enzymes that help build other molecules. For example, the number of variations of a protein molecule that is only 100 amino acids long is 22^{100}. That is a lot of options for natural selection to play with. The amino acids serve as a versatile brick from which all the proteins can be built. Similarly, nucleic acids, sugars, and fatty acids have all proven to be great bricks for building life.

When considering other monomers that might expand our range on a periodic table of life, I find myself a bit stumped for good alternatives. However, I carry a strong intrinsic bias, being made from the very bricks that I'm trying to outsmart.

Other monomer and polymer options certainly do exist, and you are surrounded by some of them every day. Plastics are a great example of polymers made from simple monomers. The most common monomer in plastics is ethene (also known as ethylene), which is a two-carbon molecule that has four hydrogen atoms (C_2H_4). Look at the bottom of any plastic bottle and you will see the LDPE/HDPE/PETE acronym for the type of polymer from which that bottle is made: low density polyethylene, high density polyethylene, and polyethylene terephthalate. Plastics are clearly quite useful when it comes to building things large and small.

The problem with plastics serving as building blocks for life parallels the problem plastics pose for us as a society: they are hard to recycle. The biopolymers of life, meanwhile, are constantly being made and destroyed in the cells of a living organism. Biology thrives by recycling monomers, and life's polymers lend themselves to easy assembly and disassembly.

Plastics do not share this property. Once plastic polymers are made, it's not easy or efficient to recycle them. At best, we tend to turn high-grade plastics into lower-grade plastics—a valiant and noble effort but still far from what biology needs from a good set of monomers.

Lacking an obvious alternative to the monomers that life as we know it uses, we can try to distill out some attributes that make for the best bricks and mortar. Here again Benner and colleagues offer some useful insight. Most of the monomers that life on Earth uses are dipoles, meaning one end of the molecule has a positive charge and one end has a negative charge. This relates back to the utility of these molecules in the polar solvent of water. But there is another, powerful advantage to using dipole monomers. Each monomer acts like a bar magnet, and as they are linked together, the positive and negative ends of each molecule in the chain come together. In amino acids, the dipole results in part from the set of bonds made between carbon, nitrogen, and oxygen in each type of amino acid. The dipole linkage makes it easy for a polymer, such as the proteins made with amino acids, to be folded and twisted into all sorts of configurations. The many configurations of proteins is, in part, what gives them the utility, either as enzymes or structures, that life needs.

As possible alternatives, with limitations, Benner and team offer up sulfonamide and phosphonamide. These two types of monomers result from replacing several of the carbon atoms in a standard amino acid chain with either sulfur, to make a sulfonamide chain, or phosphorous, to make a phosphonamide chain. Phosphorous, as we explore in the next section, is problematic because of its limited availability. However, sulfur and sulfonamides are particularly intriguing, given the abundance of sulfur on Europa's surface. Recall that Io ejects sulfur from its volcanoes, some of which eventually lands on Europa. The charged particle irradiation that bombards Europa's surface then leads to the production of an array of sulfur compounds. Perhaps if there is life on Europa, its equivalent of proteins carries a faint whiff of sulfur in a sulfonamide backbone.

At the "bricks and mortar" level of biochemistry a number of variations, large and small, might yield considerable diversity in life throughout the universe. Whether its an entirely different monomer inventory, of which we have yet to conceive, or simply the same inventory that we use on Earth, with slight changes in the amino acids and sugars, such differences would expand the classification system that weaves together all life.

BIOLOGY'S BATTERY

Along with the bricks and mortar, life also needs a basic chemical battery within each cell to power life. In chapter 9 we explored the circuitry of life. The chemical disequilibrium of an environment, or similarly the chemical potential, is akin to a battery in the environment from which life can harness the energy needed to live. But life typically cannot harness that energy directly, and it also needs to regulate the flow of that energy to the rest of the cell. To accomplish these goals, life on Earth—and potentially life elsewhere—employs specific compounds to control the flow of energy.

Within each living cell on Earth the incessant synthesis of adenosine triphosphate (ATP), and the subsequent breaking of ATP into adenosine diphosphate (ADP), serves as the local energy storage and release mechanism. Simply put, the ATP to ADP reaction involves losing a phosphate group (a phosphorous atom and three oxygen atoms) off the end of the ATP molecule. The energy released when the phosphate bond is broken is small—about 5×10^{-20} Joules per ATP molecule—but incredibly useful. For comparison, that amount of energy is one millionth, of a billionth, of a billionth of the energy stored in a size AA battery. Without ATP as biology's "in house" battery, there is no "controlled burn" of the reductants and oxidants in an environment.

All life on Earth uses ATP, but that's not to say that all life in the universe must use it too. In fact there are good reasons to avoid phosphate. Of the six elements that are most important to life—carbon, hydrogen, nitrogen, oxygen, phosphorous, and sulfur (CHNOPS)—phosphorous is the hardest to find. On Earth, phosphorous is primarily found in rocks, and it's not easily extracted. Carbon, nitrogen, and sulfur all form numerous molecules that flow through the air or dissolve in water, making them readily available. The only form of phosphorous that is typically found in water is trace amounts of phosphoric acid. Beyond that, there's very little phosphorous in the gas or liquid phase.

And yet, phosphorous is central to forming the backbones of DNA and RNA, and it's the "battery bond" in ATP. This has long been a puzzle for biologists and chemists. The solution was perhaps best summarized in a classic science article from 1987 by Harvard University professor Frank Westheimer.[10]

Westheimer concluded that phosphorous, when bound to four oxygen atoms, is particularly well-suited for the functions of life because it is trivalent and negatively charged. Trivalent means that the phosphate ion has three potential bonds to give when connecting with other compounds. This attribute makes it attractive for building long scaffolds, such as the double helix of DNA. It also means that phosphate can store energy well when bound into ATP. Phosphate's negative charge helps it interact with water and yet stay relatively stable against bonding and reacting with water instead of other compounds. In other words, phosphate is a reliable ion that can float around in the water of the cell without turning into something else or migrating through the cell membrane.

Alternatives to phosphates definitely exist and satisfy many of the conditions Westheimer laid out, but he concluded that they all had pitfalls. Citric acid, which you've likely squeezed out of an orange, could form many of the same bonds as phosphate, but those bonds would be less stable in water than the phosphate bonds. So too for acids of silicon and arsenic. This would potentially lead to bonds breaking when they are not supposed to—an undesirable attribute from the standpoint of natural selection.

At the higher pressures within ocean worlds, these problems with stability would likely persist, favoring phosphate over other options. Phosphate, despite its rarity, might be a universally loved way of doing business in biochemistry. When it comes to biology's battery, evolution may converge on the utility of phosphate, but again I am biased by the success of phosphate biochemistry here on Earth. I hope that other options exist; Mother Nature is certainly much more clever than my Earth-bound brain.

THE GREAT TREE

The parameters detailed throughout this chapter may be only the beginning of the list of options from which biology and evolution could choose. If we fast-forward decades, centuries, or even millennia, I wonder how we might be organizing information about the diversity of life in the universe (if life does indeed exist beyond Earth!). Would it be organized in a table, like the periodic table, or in a tree, similar to the Tree of Life, the way we organize our categories of life on Earth? Instead of a "Periodic Table of Life," might there be some "Great Tree of Life" that encompasses all the permutations possible in the universe?

One problem with using a tree is that, at least for life on Earth, the tree connects every organism back to a last universal common ancestor, and that ancestor represents the root for the origin of life itself. In other words, a tree implies a single, connected origin. But part of the goal of searching for life is to see if separate, independent origins exist beyond Earth. Is it possible to represent separate origins on a single "Great Tree." My rough attempt at doing that is shown in Figure 13.1. Here the trunk of the tree shows commonality of chemistry, not of origin. We start with the solvent and work our way down to the tree of life that describes life on Earth. By the time we get to the twigs representing our tree of life, the chemical differences are at the genetic level; and although they do connect back to a common ancestor, that is not required at every branching point. Indeed, the branches that form at the level of the information molecule (e.g., DNA) are probably the best we could do in terms of connecting a common origin, as it is those molecules that carry the genetic history.

Considering the option of a table, it quickly becomes clear that a two-dimensional table is not going to get the job done. Each of the parameters we discussed needs to have its own axis or dimension. Instead of a two-dimensional table, we need a multidimensional "hyper-cube." Figure 13.2 shows my effort to illustrate what this might look like if we

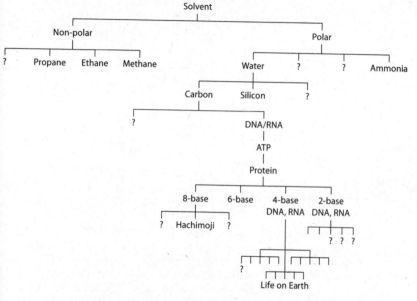

FIGURE 13.1. Great Tree of Life, an organizing tool for the diversity of life. This is my attempt to connect various possible chemistries of life through a branching tree. In this formulation, the deepest branching occurs at the level of the solvent needed for life, and the highest branches represent differences at the level of the genetic molecules used to code for life.

limit our depiction to three dimensions, one each for the solvent, the essential elements, and the information molecule. All life on Earth would occupy the point where water, carbon, and DNA-based life intersect. In reality, many more dimensions would unfold out of that cube to capture the different possible monomers (e.g., amino acids) and batteries (e.g., ATP) for life.

A thought that I like to entertain is that whatever the organizational tool for describing the diversity of life is—a table, a tree, or something else—it already exists. We humans here on planet Earth are quite young by galactic terms. We've been sentient for maybe a few million years, and we're on a planet that is only 4.6 billion years old. Meanwhile, the universe has been around for over 13.7 billion years. That is a lot of time for other civilizations to have come and gone. Perhaps, as Arthur C. Clarke,

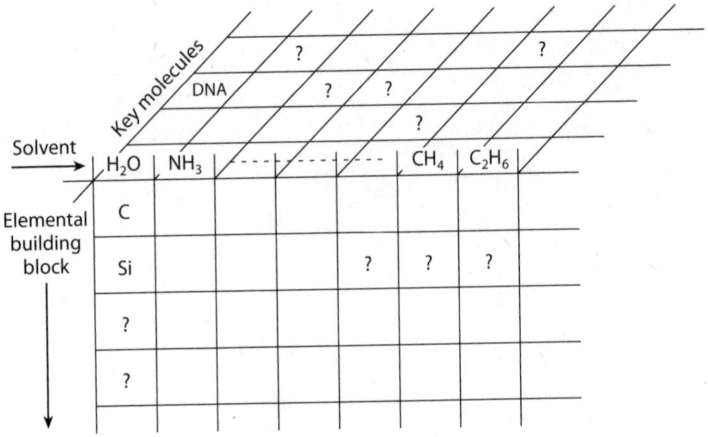

FIGURE 13.2. Periodic Table of Life, another way to organize the relationships that might exist for diverse types of life in our universe. To capture the parameters associated with biology, such a table would need far more than two dimensions. This is a rough attempt to sketch out, in three dimensions, how that information might be organized. I have limited the variables to the solvent, essential elements, and the information molecule needed for life. Life on Earth is represented by the intersection of water-, carbon-, and DNA-based life. Many more dimensions would be needed to represent, for example, different monomers and energy molecules for life.

Carl Sagan, and Isaac Asimov liked to postulate, there is some sort of *Encyclopedia Galactica* that contains all the information from all of the civilizations out there. (In today's Internet parlance, we would be more likely to refer to it as the *Wikipedia Galactica*.) Whatever the case, if there are intelligent, scientific civilizations out there that have explored their planets and nearby stars, then they too have encountered this challenge of piecing together the story of life. I wonder how they have organized their thoughts. A poster of the Periodic Table of Life, or the Great Tree of Life, may exist on the wall in some classroom out there in the stars. Who knows? Perhaps life on Earth is already on it, or perhaps our form of life is yet to be discovered.

PART IV

THE NEXT STEPS

CHAPTER 14

SEEKING SIGNS OF LIFE

Just because life exists, or has existed, on a world in our solar system or beyond, doesn't mean we'll be able to detect it. The signs of life could be quite obvious, or quite subtle.

Consider the following example. My favorite stone is marble. You've likely seen marble statues in museums, marble stairways in fancy buildings, and even entire monuments made of marble. It's a beautiful stone.

To the trained eye, marble is also a sign of life that once existed on Earth. Marble is made by subjecting limestone rocks to high temperatures and pressures, and limestone is made through years and years of sediments accumulating on ancient seafloors. The sediments that make limestone are the microscopic carbonate shells of dead organisms called foraminifera. Marble is made from the cooked and compressed shells of these tiny creatures. If you know this, then when you look at marble, you know you're likely looking at a sign of life in the past. It's a subtle sign but one that provides a useful example of the challenges we face as we search for signs of life on other worlds. If we found marble on Mars, for example, might we be able to conclude that life once existed on Mars? Perhaps. Throughout this chapter we examine the challenge of detecting life on other worlds and what that tasks entails.

To begin with, the discovery of life beyond Earth is contingent on at least three key events: the origin or seeding of life on that world, the successful colonization and persistence of life on that world, and finally, the signature that life has left of its past or present existence. Just because a place is inhabited does not mean that we have a surefire way of finding and detecting that life. On Earth, plate tectonics has recycled rocks and erased much of the fossil evidence of microbial life that existed on our planet billions of years ago.

Determining what "counts" as a strong, or even sufficient, sign of life is not easy. Biosignatures, as these signs of life are sometimes called, are the topic of much debate and discussion within the astrobiology and planetary science communities. What evidence would we need to conclude that we have found life on another world? What are the potential biosignatures that we might encounter elsewhere, and would they be convincing enough for us to conclude that we have found life? Simply put, *how* should we search for signs of life beyond Earth?

Certain types of evidence are obviously better than other types. Were we to travel directly to the oceans of Europa or Enceladus, or to the seas of Titan, and observe a creature swimming by our robotic vehicle, we might be satisfied right then and there. We saw a large swimming organism—case closed. There is life beyond Earth.

But at least for the near-term—before we get robots directly into these oceans and are limited to orbiting spacecraft and surface landers—we may have to rely on more subtle lines of evidence. Lines of evidence, which when combined, give a compelling picture of whether or not we're looking at something made by biological processes.

Composition, morphology (structure), and context are the big three—the three most important types of evidence—for any biosignature framework. Composition tells you what a thing is made of (e.g., carbon); morphology tells you what a thing looks like (e.g., microscopic cells); and context tells you where a thing came from (e.g., the ocean). Combine several measurements that satisfy these three criteria, and you may have a strong framework for detecting biosignatures and even life itself. Within each of these "big three" are many possible types of measurements, each with its pros and cons.

One of the biggest limitations is that we cannot take every instrument we want to these other worlds. It's simply not practical—there is no rocket big enough to launch a robotic spacecraft with all of the bells and whistles we would love to have to search for life. When designing such a spacecraft, we have to choose wisely, designing experiments and selecting instruments that are small enough to fit on the spacecraft but capable enough that they can get the job done. NASA has tackled the challenge of searching for life only once before, and it was in the earliest days of our exploration of the solar system. It's a mission that set the stage for our search for life, and from it we learned many important lessons that can, and will, be applied to our search for signs of life in alien oceans.

FIRST TRY, LAST TRY

There has only been one time in the history of space exploration when looking for direct signs of life was the stated target of the mission. (Technically it could count as two times, since there were two spacecraft pursuing the task.) As it turns out, this first time was also the last time.

On July 20 and September 3, 1976, the twin Mars *Viking* landers set down on the Martian surface to search for signs of life. This was a stunning, herculean achievement. NASA was only 18 years old. The first flyby of any planet occurred only 14 years earlier. In the same way that we can look back in awe at the Apollo program, which safely landed 12 humans on the Moon and returned them safely home, we should also look back in awe at the stunning achievement in robotics that the *Viking* landers represent. They landed, dug into the Martian surface, and analyzed the Martian soil in search of microbes and others creatures. The instruments on board worked beautifully, as did the cameras, the robotic arm, and all the other subsystems needed to run the robot and have it send data back to Earth.

To put this achievement into perspective, it is worth noting that just one year earlier, in the spring of 1975, a small company named Microsoft was founded. And a few months before the landers set down, another small technology company was founded on April 1, 1976. This one was named Apple. We were in our technological infancy when we landed these two vehicles on the Martian surface.

Furthermore, we were also in our infancy regarding how to think about biosignatures. We still had so much to learn about biology and the nature of life on Earth. It wasn't until 1977 that explorers discovered hydrothermal vents at the bottom of the ocean. And down in the coldest, driest place on Earth—Antarctica—scientists had just discovered microbes living within the rocks of the Transantarctic Mountains. The discovery of the structure of DNA had only been published in 1953. Perhaps most telling, though, is that it was not until a year after the touchdown of the *Viking* landers that the third major branch in our tree of life was discovered. Carl Woese and George Fox published their results for the Archaea in 1977, and by many accounts that marks the true beginning of mapping out the tree of life on Earth.[1]

The molecular and genetic techniques that are commonplace today were not available at the time of the *Viking* missions. The technique that Woese and Fox used to discover Archaea became the standard method for mapping out life on Earth and laid the foundation for mapping the human genome. The patent for that method, the polymerase chain reaction (PCR), would not be filed until 1983, long after the *Viking* landers had finished their mission.

So what, exactly, did the *Viking* landers do and what did they find? As you likely know, there were no strong signs of life detected by the landers. If the *Viking* landers had discovered strong evidence for life on Mars, you would have heard about it and every science textbook printed since the late 1970s would have had chapters dedicated to the finding.

After landing on the surface, each lander used a robotic arm to scoop up soil and transfer that soil sample to a suite of instruments. Each lander had three biology investigations: the Gas Exchange experiment, the Pyrolytic Release experiment, and the Labeled Release experiment.

The Gas Exchange experiment involved combining the soil with a liquid mixture of organics and other nutrients that any Martian microbes might eat. If there were any microbes that ate the "soup," the microbes would, the team hypothesized, release new gases into the headspace above the sample, hence the name "gas exchange." The gases were then examined, specifically for hydrogen, oxygen, nitrogen, carbon dioxide, and methane—all of which might have been the exhaled gases of Martian microbes. No evidence of gas exchange attributable to life was measured.

The Pyrolytic Release experiment exposed the soil samples to carbon monoxide (CO) and carbon dioxide (CO_2) gases that had been created on Earth and which used the heavy, radioactive form of carbon (carbon-14, ^{14}C). The hypothesis was that after exposure to sunlight and water, any organisms within the soil would consume the gases and incorporate the ^{14}C into their cells. This radioactive carbon could then be detectable by examining the chemistry of the sample. Although the experiment ran for many days, providing what most thought was sufficient time for any putative microbes to metabolize, no radioactive carbon was found during subsequent analyses of the soil. No microbes were breathing in the radioactively labeled CO and CO_2.

Finally, the Labeled Release experiment operated on a similar principle to that of the Gas Exchange experiment, and in some ways it was the inverse of the Pyrolytic Release experiment. In the Labeled Release experiment the incubation was with an organic mixture of radioactively labeled ^{14}C. If any microbes ate the organics, they would likely produce CO_2 containing the ^{14}C, and subsequent measurement of the $^{14}CO_2$ would be a very strong indicator of biological activity. Although there were some intriguing—if not confusing—results, the general agreement was that no labeled CO_2 was detected at a level that would be expected for microbes growing for a sustained period in the incubated soil.

These three experiments used the most advanced microbiology techniques of the day. To implement them on the surface of Mars was an extraordinary achievement. The only problem, in retrospect, is that all three targeted living, extant, life. In other words, in order for these experiments to generate positive results, the samples of the Martian soil would need to contain life that was alive and metabolizing. That's a tall order, even in a controlled lab on Earth. Despite what rotting food in your kitchen might indicate, it can actually be quite difficult to culture microbes.

An alternative to looking for life that is alive and growing is to look for the basic chemistry of life, regardless of whether it's alive or dead. One key instrument on the *Viking* landers, the Gas Chromatograph–Mass Spectrometer (GCMS), did exactly that and provided a pretty definitive negative result on the prospects for finding life. A GCMS is, at its core, a fancy oven connected to a system of magnets capable of sorting and detecting a wide variety of chemical compounds. Like you smelling fresh pizza, a GCMS can sense molecules popping off of samples as they are cooked in the instrument.

The GCMS instruments on both *Viking* landers found no organic compounds (no multicarbon compounds[2]), down to a detection limit of one organic molecule per one billion molecules of everything else. In other words, the GCMS was quite capable of finding that organic "needle in a haystack," and it found nothing. Given the thoroughness of the GCMS technique, it seemed conclusive that if there aren't any organics on Mars, then there can't be any carbon-based life. The ambiguity of some of the three biology experiments was, in large part, put to rest by the GCMS results: no carbon, no life.

That said, there have been a few good arguments for why the GCMS might not have seen any organics, even if they had been in the soil. The counterargument to the non-detection of organics is that the chemistry of the soil, and the heating in the oven, disguised organics as carbon dioxide.

The *Viking* GCMS instruments did, in fact, detect plenty of CO_2. The Martian atmosphere is rich in CO_2, so this was no surprise. However, there are many ways to make CO_2. One relevant example is that if you have organic compounds mixed in with very oxidizing compounds (such as chlorate salts and bleach), then you might, as the GCMS heats up, cause the organics to react with those oxidants. The net result would be that the organics are converted into CO_2, and the GCMS would measure no organics but a lot of CO_2.

Is that what happened with the *Viking* experiments on Mars? I think it could have contributed, but overall there would still have to be a small amount of organics in the soil that the *Viking* landers sampled. If there were abundant organics in the soil, the oxidants would not have been able to consume and hide them.

In the end, all four of the biology-focused experiments revealed no strong signs of biosignatures in the Martian soil. Was this evidence of no life on Mars? I don't think so. I would be somewhat surprised if Mars never had life at some point in the past. To that end, I think Mars may have some large organic molecules from ancient life hidden away in its rocks. There might even be life alive today in pockets of liquid water underground, or under glaciers and ice sheets. I think we've still got a lot of work to do on Mars, and new techniques and instruments are making each mission better than the last.

Sadly, immediately after the *Viking* missions, there was a long hiatus in Mars exploration. Some wrongly attribute that hiatus to the failure of the *Viking* missions to find life. Had the results been different, surely the discovery of life on Mars could have invigorated future exploration and maybe even catalyzed sending humans to the Martian surface. But the political reality for NASA had already been written in stone. Nixon killed the Apollo program, and the last flight to carry humans to the Moon was in 1972. NASA's budget went from nearly 4% of the national gross domestic product to only a few tenths of a percent. Decisions made in the early 1970s set NASA on a path of considerable austerity for much of the 1980s. To attribute the decline of Mars exploration to the "failed" *Viking* missions is a sloppy read of history and is poor science.

The *Viking* missions and data were great. Mars is dead. (At least its surface appears to be so.) That result still stands today and has been further corroborated by every subsequent mission that has gone to the surface. Had the *Pathfinder*, *Spirit*, *Opportunity*, *Phoenix*, or *Curiosity* missions found abundant signs of life, then one might have a valid case for saying that the *Viking* missions "failed."

Only recently have we begun to get any significant sniff of organics on Mars, and that has been by using the GCMS on the *Curiosity* rover, which landed on Mars in 2012.[3] The *Curiosity* rover will continue its rolling hike up the shoulder of Mt. Sharp in Gale Crater for many years to come. It could, at any point in that journey, stumble across an outcrop of rock that carries evidence of past life on Mars. In early 2021 a new rover will join *Curiosity* on the surface of Mars. Its task will be to hunt down interesting rocks that could contain signs of life, which will someday be rocketed back to Earth for study in the best labs, with the best instruments.

Unfortunately, in our efforts to find biosignatures on Europa, Enceladus, Titan, and the other ocean worlds, grabbing a sample and bringing it back to Earth is not an option. Those worlds are too far away, and the missions would be too complicated for any viable sample return. The one exception is flying through the plumes of Enceladus (no landing) and returning to Earth. I hope that someday we might do that; however, it could be challenging to grab enough sample worthy of making the trip.

For at least the near term, our strategy for searching for biosignatures on ocean worlds will focus on what we can do on the surfaces of these worlds with a relatively simple lander, with a relatively small payload of instruments. For that task, there are many lessons we can extract from the *Viking* experience.

LESSONS LEARNED

In our exploration of ocean worlds, we can, and must, learn from our experience seeking signs of life on Mars. Since the days of the *Viking* missions, the Mars science community has developed a strategy for studying the Martian surface and determining what—if any—regions might be habitable today and what regions might have been habitable in the past. They have also developed a framework of measurements that can be used for biosignatures but that also provides useful information even in the absence of signs of life.

This final point was one of the key lessons learned from the *Viking* experience[4]—make dual-use measurements that are good for biosignatures but also good for understanding Mars as a planet. A mission should achieve great science even in the absence of any sign of life. The GCMS measurements are a good example. The GCMS could detect and characterize organics, but it could also identify other salts and minerals in the soil. The three exchange-and-release experiments are problematic in this capacity. They did not offer much scientific value beyond the prospect of active biology results.

The framework for potential biosignatures on Mars that has been developed since the *Viking* days serves as a useful template in our search for biosignatures on the surface of any world. The key is that the framework should consist of a set of robust, redundant, and complementary measurements. Robust simply means that we understand the experiment well—it's a tried and true technique used on Earth.

Redundant measurements require that we have two or more techniques to gain confidence in the results, be they positive or negative, for biosignature detection. If one technique fails or is ambiguous, another technique should pick up the slack. For example, the GCMS on the *Viking* landers were pretty definitive for organic analyses, but it would have been nice to have another instrument to provide a second way of searching for organics. This would have increased our confidence in the non-detection of organics at the parts-per-billion level.

The *Viking* landers did have some redundancy—there were two landers, and thus duplicates of every experiment were conducted in different regions of Mars. This doesn't quite satisfy a redundancy of technique, but at least a cross comparison could be conducted between the two landers.

Having complementary measurements is one of the most critical aspects of any framework for detecting life. For example, if you combine chemistry measurements with images from a microscope, you are capturing two independent characteristics—composition and morphology— that may be indicative of life. The more complementary, independent lines of evidence, the better.

This brings us back to the "big three" themes (or types of measurements) for biosignatures at the start of this chapter: composition, morphology, and context. Together these themes address the following questions: Is the chemistry indicative of life? Is the shape and structure indicative of life? And lastly, did the sample come from a place where life could have lived (or be living today)?

Returning to the exploration of alien oceans, I'll start with the last one, context, and work backward.

When we someday land on the surfaces of these ice-covered alien oceans, our first task—after taking some stunning pictures of the landscape—will be to study the chemistry of the surface so we better understand where that material came from. Is it ice from the ocean below or is it ice full of micrometeorites and other stuff delivered from space? We need to find a sample that originates from the ocean because that's where life might live. A good example of an indicator of ice made from the ocean water below would be salts. Oceans are salty; space rocks (meteorites) are not. Finding salts likely means we're sampling material that was once part of the ocean and that any organic material also came from the ocean.

Another indicator of material coming from the ocean might be plumes of material jetting out into space. For example, someday when we set a lander down on Enceladus, I bet we send it toward the south pole and nestle it as close to the fractures and plumes as the engineers will let us. We could scoop up some fresh Enceladus snow with good confidence that it came from the ocean below.

With the context of the sample established, the next step would be to take a direct look at what's inside. The easiest and most intuitive biosignature is perhaps the one that you can judge with your eyes: its morphology. Images of the sample, images from macro- to microscopic, would be incredibly useful. Life likes to build compartments to separate itself from the environment, as we discussed in earlier chapters. Observing cell-like structure within a sample could be a helpful clue toward detecting life.

That is not to say that structures that look like cells do not arise from non-biological processes. Sure, if we grab some pond water on Earth and look at it under a microscope, we'll likely see things swimming around. But on the surface of an ocean world, we're probably talking about frozen, dead stuff. As we know from life on Earth, it can be easy to confuse microscopic structures made by geology from those made by biology. Bubbles, tubes, and other curious cell-like features in minerals and ice have all sorts of formation mechanisms that don't require life. Understanding such structures is an area of intense study for many geologists and biologists trying to find the earliest evidence for life on Earth. Nevertheless, seeing structures that we associate with life would be a very useful clue along the path toward finding life.

Complementing morphologic evidence is composition. Knowing the details of what kinds of molecules are in the sample is arguably the most definitive way to know that you've found life—short of an octopus swimming by your camera or seeing frozen shrimp in the ice.

With composition, things can get quite complex. There are a host of different compositional biosignatures, but here I'll focus on only three. To be clear, I am focusing on water- and carbon-based life because that is what we can reasonably hypothesize exists within these alien oceans.

The first example involves a search for organics and patterns in any organic compounds. It's an idea that dates back to James Lovelock pondering life on Mars.[5] If organic compounds were found in a sample, and the context establishes that they came from the ocean, we could then look for specific molecules of interest, and more broadly, the "specificity" of all the organics. A GCMS would be a good instrument for conducting this kind of analysis. The specific molecules might be compounds like amino acids or nucleotides, which make up the proteins and nucleic acids that life on Earth uses. But they could also be something completely different. By "specificity" I mean that life uses a specific set of compounds; it doesn't just piece together life with any old molecules. We spent some time on this topic when we examined the bricks and mortar of life. Life's monomers constitute a very specific set of compounds that are used to make nearly everything else. When you inventory a sample of organics from life, you see that they cluster into specific sizes that are set by the size of the monomers.

For example, your analysis might show one amino acid, or two amino acids linked together, or three linked together, and so on. You will never measure two and half amino acids; that's not how the chemistry of life works. Contrast this with organics made by random chemical and geological processes. While there certainly can be some specificity and patterns to those processes, they are typically more likely to generate a random mix of organic molecules not based on selective monomers. These two patterns, the "specific" and the "non-specific," serve as a signpost for distinguishing life from non-life.

An added, and very useful attribute of this metric is that it can be generalized to any kind of biochemistry. Life using something other than DNA, RNA, and proteins will still likely use specific monomers because the functions and organization of life necessitates selection of the best-suited compounds. In other words, life and biological processes are specific, not random, and we can measure the difference between these two scenarios with instruments like a GCMS.

The next example of a compositional biosignature may also be universal to all organic life, and possibly even to other weird forms of life. It involves chirality, which is a fancy word for handedness. A chiral molecule, in the simplest form, is one of a pair, where nature has made two versions of the same molecule, each the mirror image of the other. It's a subtle but important difference in the molecular structure.

For example, consider looking at yourself in the mirror while wearing a shirt with a breast pocket. Shirts are usually made with pockets on the left side, but in the mirror version of you the pocket is on the right side. If you had two identical shirts, one with the pocket on the left and one with the pocket on the right, you could be said to have a pair of chiral shirts, one left-handed and one right-handed.

In chemistry, the equivalent of the shirt pocket is usually some oxygen, nitrogen, and/or hydrogen atoms dangling off the molecule. That group of atoms breaks the symmetry and makes the molecule chiral—there's a left-handed and right-handed version.[6]

Chirality sounds like a relatively small attribute, but it turns out that there are big consequences. Many of the monomers of life are chiral; in nature you find both versions, but in life you only find one. On Earth life uses only left-handed amino acids and right-handed sugars. There is not really any conclusive reason why evolution chose this handedness, but once established, it probably stuck. The reason it stuck is because monomers of one hand can only link with molecules of the same handedness. Polymers, such as proteins and polysaccharides, are therefore only made from monomers of one handedness, left or right.

To understand why this needs to be the case, imagine building a spiral staircase that goes several stories up. When you build the first floor of stairs, you have to decide in which direction—clockwise or counterclockwise—the spiral staircase will ascend. Let's imagine that you chose clockwise. Now when you go to build the second story, you clearly have to stick to that direction because a counterclockwise staircase will not connect to the clockwise staircase on the first floor. Chiral molecules linking together to make large molecules, like proteins and polysaccharides, operate in a similar fashion. Once the chirality is determined, all the linking monomers have to follow suit. Early on in the history of life on Earth this selection was made. It is entirely possible that right-handed amino acids and left-handed sugars could have "won," but they didn't and we're now locked in to the system we have.

As a biosignature, chirality is quite useful. A sample full of organics not made by life will have roughly equal numbers of left-handed and right-handed molecules. A sample containing organics made by life, however, will have only one handedness for each kind of molecule. On Earth, a biological sample will have only left-handed amino acids and right-handed sugars.

There are some nuances to this—it's never a perfect all-or-nothing. Various processes in the environment, such as heating and radiation, can affect the strength of the chiral signature. Molecules of one handedness can be converted into the opposite handedness, transforming a strong one-handed biosignature into a more ambiguous mixture that's closer to a 50-50 split in handedness. Nevertheless, chirality is generally considered a pretty robust biosignature, especially if you understand the context of the sample, such that you can estimate how much racemization, or equalization of handedness, has occurred.

The third and last example of compositional biosignatures involves the measurement of carbon isotopes. Recall that isotopes of a given element are simply the variations on the number of neutrons carried by the atom. The most abundant isotope of carbon in the universe has six neutrons to go with the six protons. Carbon has one stable isotope, carbon-13 (^{13}C), that has seven neutrons. Carbon-14, which we discussed in the context of the *Viking* experiments, has eight neutrons, but it decays to nitrogen (a neutron turns into a proton) with the relatively short half-life of 5,700 years.

Carbon-13 constitutes just over 1% of the carbon in the universe. But in biology—in the organics of life—^{13}C is even more rare. It constitutes just 0.01%–0.1% of the carbon used to build life. The exact reason for this difference varies depending on whether you're looking at plants, animals, or microbes. But at the heart of it is the difference in mass and size of the two isotopes. In many biological processes, when carbon is being metabolized and incorporated into the cell, it takes less energy to use ^{12}C instead of ^{13}C. For that reason, as an organism grows, it ends up preferentially building itself with ^{12}C and rejecting ^{13}C. Life is, truly, enlightened, at least relative to the carbon in its environment.

What this means from a biosignatures standpoint is that if you measure the ratio of ^{13}C to ^{12}C in a sample, you can get a pretty good sense of whether it came from life or non-life. A biological sample will be depleted in ^{13}C, whereas a non-biological sample will typically have a concentration of ^{13}C comparable to that of other rocks and compounds in the environment that are not linked to life.

One limitation of this powerful technique is that you have to know the isotopic ratio of the carbon in your environment in order to understand what the fractionation of carbon means. If, for some reason, you live on a planet that has more ^{13}C than the Earth does, then that will affect the relative ratio of the isotopes in different materials. On Earth, geochemists have made these measurements in places all around the planet, and thus we have a good context for the isotopic depletion of ^{13}C in biological organics. On Europa, Enceladus, and Titan, we would have to be more cautious with our interpretation of any isotopic signature, in part because we would likely not have many measurements from a variety of sites on those worlds. We would lack strong context for the interpretation. Despite this potential shortcoming, isotopic fractionation is still a compelling biosignature measurement and may also serve as a universal biosignature, at least for carbon-based life. It is reasonable to expect that the same energetic driver for preferentially using ^{12}C would prevail in alien biochemistry.

To conclude, the above suite of measurements constitutes the beginnings of a robust, redundant, and complementary set of potential biosignatures that could be used to search for life on the icy surfaces of ocean worlds. Were we to make all of these measurements and they indicated life, we would be well on our way to confidently claiming that life had been discovered beyond Earth. It is a framework that has been developed through decades of studying life on Earth and through attempting to find signs of life on Mars.

Over the past decade many teams of scientists have further refined and improved our search strategy. In 2016, NASA convened a formal team of scientists to define the goals and strategy for a lander that will hopefully someday go to the surface of Europa. That strategy carried many of the lessons learned from *Viking* and all the discoveries in biology and geology that have been made since that mission. As we approach the fiftieth anniversary of the *Viking* missions (2026), hopefully we will be well on our way to getting such a mission to the launchpad.

———

In closing, it is worth noting that on at least one occasion, NASA successfully detected life with a robotic spacecraft that was designed to study distant worlds. That search did not use any of the techniques described in this chapter, nor did it find life on an ocean world. But it did prove that we could, indeed, find life.

In 1990, the *Galileo* spacecraft flew by the Earth on its way to Jupiter. To test out the instrumentation on the spacecraft, a few observations were made of our home planet. Although the instruments were designed to study Jupiter and its moons, this flyby of Earth provided a serendipitous opportunity to see if data from *Galileo* could reveal enough information about Earth to show that it was, in fact, inhabited.

The idea to do this "proof of concept" was conceived by Carl Sagan, Bob Carlson, and Reid Thompson.[7] The plan that the team put together was relatively straightforward: as *Galileo* flew by Earth, images, spectra, and radio frequency signals would all be collected.

Not too surprisingly, *Galileo* detected radio emissions from Earth that were best explained by a non-naturally occurring source, i.e., radio broadcasts from technology created by intelligent life. Images from *Galileo* also revealed a planet with continents and oceans. On some of these continents, the images showed a sharp "red edge," or drop in the intensity of color near about 0.7 microns in wavelength. That edge in the spectra had one very good explanation: plants and photosynthesizing microbes. Photosynthetic life absorbs sunlight in this region, and thus the camera filter for the red region was quite dark compared to other filters. The "red edge" is a strong indication of vegetation.

Lastly, data from *Galileo*'s spectrometers revealed a curious characteristic of Earth's atmosphere: it was (and is) in a state of chemical disequilibrium. The instruments detected ozone and oxygen, two very powerful oxidants. They also detected methane, a strong reductant. On a world without life, methane and oxygen could not be sustained in an atmosphere for any significant amount of time—methane is destroyed by sunlight, and oxygen and methane can combine to form carbon dioxide and water. The observation of these two molecules in Earth's atmosphere was a sign of waste gases from life. Trees and plants exhaling oxygen, and among other contributions, the methane source was "flatulence from domesticated ruminants." That was Sagan's fancy way of saying "gas from cows."

As we prepare to explore alien oceans beyond Earth, we have to consider the variety of ways past or present life may show signs of its existence in the near-surface ice of these worlds. From the subtle magnificence of marble to flapping fish in front of a submersible, all options—all possible biosignatures—must be part of our framework. Searching for and verifying the existence of life (past or present) is a difficult task and one that mandates the utmost caution when it comes to the interpretation of the data. Sagan and colleagues put it best, when they advised, in the study described above, that as we move forward with the search for life, we must rule out all other interpretations, as "life is the hypothesis of last resort."

CHAPTER 15

A NEW AGE OF OCEAN EXPLORATION

In 1997 I was a young college student obsessed with aliens. That same year, results from the *Galileo* mission were starting to make a strong case for an ocean beneath Europa's icy surface. NASA even released an image with an artistic vision of how we might someday send a melt probe through the ice and into the ocean. The far-off date for that melt-probe was listed as the year 2009.

Well, 2009 came and went. No melt probe made it into Europa's ocean. There was no lander to sample the surface. There wasn't even a spacecraft sent to fly by and take a closer look.

NASA had convened many studies for all sorts of missions to Europa, but the moment things looked promising, NASA, the scientific community, or politicians, would pull the plug. By 2009 I was a scientist at the Jet Propulsion Laboratory, and I had already spent three years as part of a team, led by my colleagues Professor Ron Greeley, Dr. Bob Pappalardo, and Dr. Louise Prockter, working on those mission studies. It would take another five years before things started looking up for getting a mission to the launch pad. As I write, in the year 2019, we are moving forward with building a spacecraft that will fly by Europa dozens of times. That mission will not get to the launchpad before 2023, and probably will not return its first close-up image of Europa until the late 2020s. All told it will likely be 30 years after NASA released that inspiring image of Europa's ocean when we finally get new images of Europa's surface. However, that mission, called Europa *Clipper*, has no lander or submersible. Those missions could take an additional 30 years each, at the pace we're going.

This business is not for the faint of heart.

But before I detail the path ahead for exploring alien oceans beyond Earth, I want to take a moment to consider progress on Earth, and how we are doing with the exploration of our own ocean. This is important because we need to take care of our ocean, and there are still many "alien" regions yet to be explored. In addition, the exploration of our ocean and the alien oceans beyond Earth are technologically intertwined. Long before we drove robotic vehicles on Mars, we tested them out in the Mars-like deserts of Earth. So too for the alien oceans: before we send a robotic submersible into the oceans of Europa, Enceladus, or Titan, we'll need to develop and test that technology in the depths of Earth's ocean. We're still a long way off from anything small enough and light enough to fit on top of a rocket, but the development of these technologies has the wonderful benefit of advancing our capability to explore Earth's ocean while also building toward exploring oceans beyond Earth. So where are we with the technologies we use to explore Earth's ocean?

"YOU'RE GONNA NEED A BIGGER BOAT"

The 1960s saw the dawn of the space age and incredible progress in a short time. Within a decade, we went from a nascent space program to landing humans on the moon and safely returning them home. Along with exploring the worlds above, the 1960s were also supposed to be a pioneering decade for exploring the depths below. Our ocean presented a vast and exciting frontier with many discoveries yet to be made. In 1960, US Navy Lieutenant Don Walsh and Swiss engineer Jacques Piccard, made the first human-piloted dive to the deepest depths of our ocean. They plummeted 7 miles (11 km) to the bottom of the Mariana Trench in the Pacific Ocean. Many hoped that would be the beginning of a committed effort to map out every corner of our ocean.

And yet, 60 years later, much of the seafloor still remains unknown.

Looking back on those early discoveries and the pioneering work in our own ocean, one can't help but wonder why much of the ocean still lies in darkness. In an age when we've mapped nearly every inch of land and mapped many other planets, why have we fallen so short in our efforts to map and explore our ocean in detail?

For one, it's hard. Unlike with the surface of the Earth, or even the surfaces of other planets, images taken from a distance are of little help. We can map planets at a distance with cameras on satellites. The images can be high resolution and can be collected with rapid efficiency.

But water—the ocean—confounds any known method of visual penetration; you can't peer through the surface to the seafloor. To get high resolution, meter- to centimeter-scale images of our ocean's floor, you have to get up close with a submersible and lights.

Some might rightfully argue that we need to spend more money on ocean exploration, and I certainly agree, but that fails to address a larger problem. When it comes to money, the oil and gas industries—not to mention the mining industries—spend far more every year than any federally funded research agency like the National Oceanographic and Atmospheric Administration (NOAA), the National Science Foundation (NSF), or NASA. More research dollars could help, but industry already has the necessary incentive. So what is wrong?

The heading of this section—"You're gonna need a bigger boat"—is a classic line from the 1975 film *Jaws*: after seeing the size of the shark, the team realizes that their boat is too small. In some ways, the need for a bigger boat is also at the root of our limited capability for exploring our ocean.

I think, as do several colleagues in the oceanographic community,[1] that we developed some bad habits in the early days of deep ocean exploration. Consider the *Challenger* expedition discussed briefly in chapter 1. It was only possible because the British Royal Navy loaned one of its ships to science. The *Trieste*, similarly, was owned by the US Navy and was even accompanied by large Navy vessels providing surface support. Fast-forward to the present day, and much of our oceanographic progress has come from scientists using ships inherited from naval programs. Many of today's research vessels were yesterday's military ships.

Why is this bad? Isn't that a great use of retired military ships? Wouldn't progress have been slower without those ships and investments? Yes, and absolutely yes.

The use of repurposed military ships has tremendously benefitted ocean science. However, having access to these large ships meant that scientists and engineers chose to build huge instruments and submersibles; they had no incentive to make small, fast, and light systems for exploring the ocean. The result is that we've built our tools to fit the big toolbox, with sample collection devices and science instruments as large, heavy, and capable as the ships we inherited. Why build something small and light when you have a massive Navy-style crane on your ship? Why bother making measurements of rocks directly at the bottom of the ocean, when you have a powerful winch that can drop and recover a massive bucket that dredges across the seafloor?

Another result of the "big ship, big science" approach is that our program of ocean exploration has been tied to a relatively small number of large research vessels that are expensive to operate. The science community is constantly scrambling to get time on these ships; and when you do get time, you are desperate to get as much done as possible. Thus you use those big tools to get as many samples and dives in as possible. It's a feedback loop where big, rare ships cultivate big, ambitious, and expensive science.

This is not intrinsically a bad thing, but it is a limiting factor for exploring all of our ocean's seafloor since there are few ships. Given how little we've explored our ocean, there's a lot of room for exciting discoveries to be made with relatively small and simple systems: a few cameras, sensors, and a modest sample collection device can return a lot of science in a place no one has ever seen before. But most ocean science programs haven't invested in these kinds of tools; so even if you get time on a smaller ship—say from a foundation or a billionaire's boat—you probably don't have a good set of tools that can fit on that boat. The vast majority of instruments and submersibles (robotic or human-occupied) weigh thousands of pounds and require big cranes and winches to deploy and recover.

Contrast this with the space program.[2] Our efforts to explore space have always been incentivized by a need for smaller, lower mass, and lower power robots and instruments.

Why? In the space world we suffer from what many refer to as the "tyranny of the rocket equation."[3] Simply put, what goes up takes fuel— and if fuel is going up, it also takes fuel to lift the fuel. The most cleverly engineered, efficient rocket is still about 85% fuel by mass. Everything else, from the structure of the rocket to the payload sitting atop it (collectively called the "dry mass"), is only 15% of the total mass on the launchpad.

To put this in perspective, consider your car. A gallon of gasoline weighs about 6.2 pounds (2.8 kg), and a typical gas tank holds 12–15 gallons, which is 75–93 pounds of gasoline. If cars were subject to the rocket equation, then the total mass of your car and anything in it would be limited to, at best, 13–16 pounds. Most cars weigh 3,200–4,200 pounds. The "dry mass" of a car is much more than the "wet mass" of the fuel it carries. When it comes to going straight up into space, the rocket equation really is a tyrant. The laws of physics are kind to horizontal rolling motion (what a car does), but very unkind to vertical motion against gravity. That is ultimately why the rocket equation is so harsh: you are trying to move directly up and against gravity (or, if landing, trying to protect yourself from gravity and stop yourself from crashing).

The rocket equation has forced us in the world of space exploration to always be thinking about ways to make our robotic vehicles and instruments smaller and lighter. It is a frustrating but useful force. We have to tighten our belts and think hard about exactly which measurements we want to make and which instruments to send. The rocket equation inspires creativity and innovation.

Consider what has been done on the surface of Mars. The *Curiosity* rover[4] on Mars has a mass of just over 1,980 pounds (900 kg), of which 165 pounds (75 kg) are science instruments. The total payload mass, which includes the cruise stage to get to Mars and all the rockets and components needed to land, comes in at 8,463 pounds (3,893 kg). So already we are looking at a ratio of 165 pounds of science instruments to 8,463 pounds of total payload mass. That ratio is 0.019, or just under 2%.

But all of this is riding atop a rocket. How much did that rocket and all its fuel weigh? The *Curiosity* rover launched on an Atlas V 541 rocket which, when fully fueled and with the payload attached, weighed

1.17 million pounds (531,000 kg). Let me repeat that with some zeros in place: 1,170,000 pounds! We landed 165 pounds of science instruments on Mars but it required 1,169,835 pounds of additional mass to get it there and to make it all work once on the surface. The ratio of instrument mass to total rocket mass is only a bit more than one one-hundredth of a percent of the total mass (0.014%).

Again, returning to the car example, imagine if the "science payload" that you were transporting were groceries, and you were constrained by the rocket equation and the numbers above for getting to Mars. If you had a 15-gallon tank, then the mass of your gas would be about 93 pounds; everything else would need to add up to 16 pounds, which is a total of 109 pounds. Taking 0.014% of that leaves you with only 0.015 pounds of groceries! One average-sized grape weighs about 0.015 pounds. That is what you get to bring home on a full tank of gas! As depressing as this analogy sounds, the brightest rocket scientists in the world would be thrilled to get that little grape back home.

Welcome to the world of spaceflight and the tyranny of the rocket equation. This is the universe in which we live.

What this all means is that scientists and engineers in the realm of space exploration have been incentivized to innovate in a way that the oceanographic community has not. We have been trained to make things small, light, fast, and efficient, and we savor every little sample we analyze and every bit of data we get back. In addition, we rarely bring back samples; the rocket equation makes it too darn hard to bring samples back to Earth. In most cases, it makes more sense to send a suite of small instruments to perform the analyses on-site.[5]

It is my hope that by the year 2050 we are operating missions capable of melting through the icy shells of worlds like Europa and Enceladus. To do that, we have to figure out how to make submersibles, drills, melt probes, and instruments small and light enough to ride atop a rocket. These robotic vehicles will have to be highly capable and also incredibly clean; we do not want any microbial hitchhikers on our spacecraft. We do not want to transport Earth life to these alien oceans. If Europa has life, we should protect that life for the Europans. So too for any

Enceladeans, Titans, and so on. Given all of these challenges, we are going to have to be very clever.

Before we send these probes to distant oceans, we will need to test them on planet Earth. A prime place for honing our exploration skills will be Earth's ocean and ice sheets. To get into the ocean worlds beyond Earth, we will have to develop new tools for exploring the ocean on our home planet. Our exploration beyond Earth will also greatly benefit our exploration of planet Earth.

It is a beautiful win-win situation that will open a new age of ocean exploration.

UNCHARTED WATERS

The *Galileo* spacecraft deliberately plunged to a crushing death in Jupiter's atmosphere on September 21, 2003. The *Cassini* spacecraft also died a deliberately crushing death, barreling into Saturn on September 15, 2017. Both spacecraft completed pioneering missions that were extended far beyond their original plan, and they were sacrificed to protect any possible inhabitants of Europa and Enceladus (making sure that Europa stays safe for the Europans, and Enceladus stays safe for the Enceladeans!).[6] With the end of the *Cassini* mission, our efforts to explore alien oceans have gone dark, and they will likely remain so for at least a decade.

The next step of exploration is a challenging one that will take a long time. This is not because of the technological challenges—those are small compared to the sociological and political challenges inherent in funding NASA missions. The missions that NASA chooses to pursue are largely determined by the planetary science community—which consists of about 2,000 people, but only a fraction are interested in the search for life. The only way to move forward with a robust program to search for life is if the public takes a more vested interest in the details of NASA's missions.

In the United States, taxpayers are NASA's sole source of funding and are its ultimate customer. I have given many talks around the country, and one theme consistently emerges: the people who pay for NASA want it to do more when it comes to the search for life. They want to see more robots scouring the cliffs of Mars, touching down on the surface of Europa, coasting in through the atmosphere of Titan, and careening across the plumes of Enceladus. Indeed, many believe that NASA *is* moving forward aggressively with such plans.[7] Dozens of studies are conducted every year, and the news often reports on them because the studies come with exciting graphics and are great for the imagination. But when it comes to actual missions that advance the search for life in alien oceans, there are very few that are truly moving forward.

The Europa *Clipper* is one of those missions, and it is a fantastic mission. *Clipper* will return astonishing images of Europa's surface, spectra across many different wavelengths, radar profiles that may allow us to "see" down into the ice, and magnetic field measurements that will enable us to continue sensing the ocean through its induction response. In addition, the spacecraft will carry instruments for studying the charged particles around Europa and measuring the temperature of Europa's surface; it has two different kinds of mass spectrometers that will be particularly useful if we find plumes of water erupting out of Europa's ice shell. These two spectrometers will be our best chance for detecting any molecules that could give hints of life on Europa. With any luck, *Clipper* will launch from Kennedy Space Center in the early-to-mid-2020s and arrive at Jupiter in the late 2020s.

Nearly simultaneously, the European Space Agency is building a spacecraft for their *Jupiter Icy Moons Explorer* (JUICE) mission. JUICE will tour much of the Jovian system and send back data from flybys of the large moons before it goes into orbit around Ganymede. Once in orbit, JUICE will provide a close-up view of Ganymede. The spacecraft will have many of the same instruments that *Clipper* is carrying, and thus it will be able to determine the composition and structure of Ganymede's ice shell in much greater detail than ever before. By merit of being in orbit, JUICE will be able to map and monitor the gravity and magnetic fields of Ganymede with great precision and accuracy.

From the combined dataset of JUICE and *Clipper*, the scientific community will be able to peel away the layers on each of these moons and learn about their physics, chemistry, and geology. These two spacecraft will be doing a beautiful orbital ballet while sending back information that will advance our understanding of exactly how habitable these distant alien oceans might be.

Beyond the *Clipper* mission, NASA has not committed to any further exploration of Europa. Since the early 2000s, sending a lander to Europa's surface has been a high priority for science. In fact, the 2003 report from the National Academies of Sciences, which outlines the science and exploration priorities for the solar system, discussed not one but *two* landers.[8] However, the 2011 version of the report cited no landers.[9] Largely for this reason, the planetary science community and NASA decided not to move forward with a lander mission, even though significant funding was made available.[10] Many other factors come into play with these decisions, including cost and schedule overruns on other missions.

In many ways, one of the key bottlenecks right now is skilled labor. According to the Office of Inspector General report, NASA and its subcontractors do not have a large enough skilled work force to complete the projects to which it is already committed. This I find depressing. Thousands of people would love to work for NASA to help make missions of all kinds possible. Surely, the talent exists for NASA to grow and do more. Public support and excitement for the agency seems to be rising—from street corner shops to department stores, the NASA logo is everywhere.

If anything, my sense of public opinion is that there is a hunger for more, for NASA and the new generation of space companies to push the frontier in ways that will help us take care of our home planet, while also pioneering grand new discoveries on distant worlds. Humanity is hungry for unifying, bold achievements that bring us together, rather than tear us apart.

Related to this sentiment, at the Jet Propulsion Laboratory in Pasadena, California, the motto is "Dare Mighty Things." It's a phrase adopted from a speech by Teddy Roosevelt[11] and popularized by former JPL director, Dr. Charles Elachi. From the standpoint of solar system exploration, few endeavors are more daring, bold, and unifying than the search for life. At the heart of this search is an intrinsic, primordial desire to understand who we are, where we came from, and why we are here.

Biology is the stuff of us, and the cosmos may have answers as to the universal role of biology.

To that end, if we were able to move forward with the exploration of distant oceans and the search for life, the path is quite clear.

For Europa, the next step after the *Clipper* mission would be to put a lander on the surface. The lander would use a robotic arm to excavate into the surface and scoop up samples that would be fed into a suite of instruments designed to search for biosignatures. This mission concept was studied in detail by a team of 21 scientists, including me, in 2016.[12] The mission was developed to the point of *almost* moving into the more rigorous phases of mission design, before it was tabled for many of the reasons cited above. It is unclear if and when the Europa lander might move forward. Many other missions may take priority over landing on Europa, and thus I fear that we are looking at a future in which we do not land until the mid-2040s.

After landing on Europa and determining whether or not there are signs of life in the ice, we would then send a probe to reach the ocean below. Even if no life was found in the ice, we would likely want to get directly into the ocean with a submersible, both to look more closely for life and to better understand the ocean itself. Keep in mind that even in the absence of any life, the chance to study the physics and chemistry of a second ocean would greatly advance our understanding of how oceans work. Planetary science has long been a field motivated by the desire to understand planetary processes across a variety of conditions. From plate tectonics on Mars, to the greenhouse effect on Venus, we have learned a lot about how planet Earth works by studying other planets. With the exploration of Europa's ocean, we could open a new field of comparative oceanography, advancing our understanding of how oceans work.

A robotic spacecraft to get through the ice and into the ocean would need some combination of a mechanical drill and hot tip to melt through the ice.[13] Like a sharpened pencil, the probe might take weeks, months, even years to get through kilometers of ice before it reaches the ocean. The probe would likely need nuclear power, as solar power and batteries would not be sufficient. As it made its way through the ice, the probe would need to trail a fiber-optic cable behind it, and perhaps some acoustic relay devices so data from the probe could be sent to the surface and back to Earth.

Once in the ocean the probe would release an autonomous underwater vehicle (AUV) capable of navigating the ocean largely on its own. This robot would not be controlled by engineers back on Earth using joysticks. It would need onboard intelligence to navigate the complex environment.

If life is in the ocean, it may be that the AUV finds it burrowing into the bottom of the ice shell, feeding off the chemistry of the ice. If not there, then the AUV would swim down to the seafloor, all the while sniffing around for any hydrothermal vents. I love to imagine this brave little robot following a chemically rich plume, making its way through black "smoky" clouds, only to discover a tower full of life as the smoke clears. Alien forms of shrimp, squid, crabs, and tube-worms swirl around the chimney. Life not only existing, but thriving in this deep ocean.

But then again, there could be nothing. Not even the tiniest speck of life. Each outcome is profound.

Importantly, the lander and probe design would be equally useful on Enceladus, Titan, Triton, and the other ocean worlds. For Enceladus, however, our first step would be to send a much smaller version of the *Clipper* spacecraft to fly by and sample the plumes in detail, searching for biosignatures as it sweeps up plume particles. This spacecraft would also likely image the surface at high resolution and pave the way for a lander. If I had my way, I'd build two nearly identical landers at the same time, one for Europa and one for Enceladus. We'd launch the missions in tandem, beginning a remarkable age of ocean exploration.

On Titan, the job of getting to the surface is made easier by the atmosphere. The *Huygens* probe demonstrated that parachutes work very well on Titan. For this reason, we do not need complex thrusters and landing systems that spacecraft to the surfaces of Europa and Enceladus would require. In addition, one could design a lander that targets a large sea or lake, which would ensure a flat, soft landing site. Instead of touching down on a hard surface, the vehicle would splash down into a liquid. As long as the lander was designed to float upright, it would be safe.

In 2009, Dr. Ellen Stofan, then of Proxemy Research and now at the Smithsonian Air and Space Museum, proposed such a mission, called the Titan Mare Explorer (TiME). It made it through NASA's first round of competition but, sadly, not through to final selection.

Thankfully, in 2017 there was another chance to propose a mission to Titan. This time the competition allowed for a larger budget and more science. This time an incredibly exciting and bold option for Titan was proposed: send in a rotorcraft similar to the quadcopter drones that are used by professionals and hobbyists for all sorts of applications on Earth.

It was a crazy idea, but it made a lot of sense. The gravity on Titan is just under 15% of the Earth's, and Titan's thick atmosphere is perfect for flying. My colleague Dr. Ralph Lorenz, at the Johns Hopkins Applied Physics Lab, has even done the calculations to show that, in theory, humans with prosthetic wings could easily fly on Titan (assuming of course, that they were protected from the toxic atmosphere and brutally cold temperatures).

Dr. Elizabeth Turtle (who goes by Zibi), of the Johns Hopkins Applied Physics Lab, led the effort to get this robotic rotorcraft out to Titan. She named the mission "Dragonfly" because the rotorcraft will buzz around Titan and hop to various sites on its surface in a manner similar to the way dragonflies move.

When Zibi and her husband (the above-mentioned Ralph) first asked me to join their team, I was honored but I also thought they were crazy. NASA never selects such bold, daring, and risky missions. We talked through the details over lunch, looking out at the ocean near Woods Hole, Massachusetts. It was a fitting view for pondering the exploration of alien oceans.

In June 2019, much to the delight—and perhaps astonishment—of our team, the Dragonfly mission was selected by NASA. If all goes well, a robotic spacecraft will parachute into Titan's atmosphere in the year 2035. It will fire up its propellers and land softly on Titan's surface. Once there, it will use a suite of instruments to look for biosignatures on Titan and investigate the geology, geochemistry, and geophysics of its surface and subsurface. After completing analyses at the initial landing site, the spacecraft will then fly to another site, and another, and another, over the course of roughly three years. It is an incredibly exciting mission that pushes the frontier of exploration in exactly the way NASA was chartered to do. I cannot wait to see Titan up close!

Moving out farther in the solar system, the next step for the moons of Neptune and Uranus is to get a mission out there to study these worlds in much greater detail. It has been over 30 years since *Voyager 2* flew by Uranus and Neptune. I am hopeful that a mission resembling the *New Horizons* spacecraft, or a simplified version of the *Clipper* spacecraft, could be sent to study one, or both, of these worlds in the coming decades. Such a mission would potentially reveal that the moons of Neptune and Uranus could be habitable, after which we would send a spacecraft to land on the surface to directly look for any signs of life.

The missions I have described above may take decades, or even a century, to complete. Again, this business is not for the faint of heart. Building robotic spacecraft to explore the solar system is the modern analog of building cathedrals. Cathedrals took decades to centuries to build, and many generations of builders. They are grand achievements that took a vision from leaders to maintain progress over decades. Rarely did anyone who laid the first stone see the last stone set in place. Our missions to explore the solar system are the cathedrals of our day. They take time, commitment, and vision.

———

To conclude, we cycle back to the beginning—of this book and of solar system exploration. In chapter 1 we covered the meticulous observations of Galileo Galilei and how the discovery of Jupiter's moons helped put an end to Aristotelian cosmology, opening the door for the Copernican Revolution. Of all the stunning images of our solar system, including those sent back by spacecraft, my favorite images are still those hand-drawn sketches Galileo made of Jupiter on those cold January nights in 1610 (Figure 15.1).[14]

I like to imagine what it was like for Galileo to make that discovery. Seeing something, discovering something, realizing something revolutionary for the first time in the history of humankind.

FIGURE 15.1. On January 13, 1610, Galileo Galilei turned his telescope to the night sky and saw Jupiter and four bright objects arranged along a rough line. His sketch, shown here, is a record of the first time he saw all four large moons. Galileo initially thought the bright objects were distant stars; but night after night he saw the objects revolve around Jupiter, leading him to conclude that they must be moons. (Reprinted from Galilei, G. (1989). Sidereus Nuncius. Chicago: University of Chicago Press)

In today's modern age of robotic exploration, I also like to think about the tiny electronic brain of a robotic spacecraft that might, unknowingly, make another profound discovery that could change our understanding of our place in the universe.

Imagine a small but capable lander on the surface of Europa, collecting samples and sending back data. The signal from the lander could take 45 minutes to arrive on Earth, perhaps longer. The signal—carried on a series of rhythmic waves tapping away at the electromagnetic fabric of our solar system—would encode a series of 0's and 1's. These in turn would be translated into a wealth of chemical information from a sample collected and analyzed on Europa's surface. The bits soaring through space could carry the answer to one of humanity's oldest questions: Are we alone?

In some ways, the robotic spacecraft sending the signal would already know the answer. The first person or thing to know whether or not life exists beyond Earth could be that humble little robot. For many minutes, it would be alone with this profound knowledge. That little brain, created by human hands, might know whether biology is a rare phenomenon, or if life arises wherever the conditions are right. It might know whether we live in a biological universe, or one in which life on Earth is a biological singularity.

Part of what excites me about the time in which we live is that for the first time in the history of humanity, we have the tools and technology to potentially answer this profound question that humans have been pondering for centuries, if not millennia: Are we alone? My hope is that centuries from now, our descendants will look back at this point in history with the same sense of awe with which we regard Galileo and the Copernican Revolution. They will look back at our time and say that it was then, *it was during this time in history*, when we built the spacecraft and we did the exploration that (potentially) brought the universe to life.

The discovery of life beyond Earth, or conversely the discovery that it does not exist anywhere else, is as profound a shift in our framework of the cosmos as is moving the Earth from the center of the universe to being just one of many planets, orbiting an average star, in a universe full of stars.

Perhaps we are the only ones. Perhaps the origin of life is hard, and life is rare. Or perhaps we live in a universe teeming with life—a biological universe of incredible diversity across planets, moons, stars, and galaxies. Perhaps our tree of life—the singular, center of biology as we know it—is revealed to be but a tiny twig, on a tiny branch, joined to a vast and grand tree of life connecting the beauty of all life in the known universe.

Looking up at the night sky, seeing Jupiter as a bright point of light above the horizon, I can't help but wonder whether our return to that beautiful planet and its magnificent moons will once again catalyze a scientific revolution in our understanding of our place in the universe.

Europa, and the many alien oceans of our solar system, await.

NOTES

CHAPTER 1. OCEAN WORLDS ON EARTH AND BEYOND

1. Sagan, C. (1997). Pale blue dot: A vision of the human future in space. Random House Digital, Inc.

2. The Hunt–Lenox Globe (ca. 1510, Rare Book Division, New York Public Library, New York, NY) includes the phrase "HC SVNT DRACONES" directly below the equator, on the southeast coast of Asia.

3. The painting *Alexander the Great under Water* (ca. 1400–1410, artist unknown, created in Bavaria, Germany, J. Paul Getty Museum, Los Angeles, CA, Ms. 33, fol. 220v) depicts the ruler and explorer inside an early version of a diving bell.

4. Beebe, W. (1934). Half Mile Down. New York: Harcourt, Brace and Company, p. 344.

5. Beebe, Half mile down, 225.

6. The *Trieste* is usually described as a bathyscaphe, which translates from Greek to "deep ship." Bathyscaphes are typically described as a type of submersible in which the vessel containing the passengers is made buoyant by a large container of a buoyant liquid, such as gasoline. Conceptually, bathyscaphes were originally conceived of as the deep ocean analog of hot air balloons. This is not surprising considering the history of invention: the inventor of the bathyscaphe concept was Auguste Piccard, who had become famous in the 1920s and 1930s for his innovation of hot air balloons, which carried him ten miles above Earth's surface, where he could begin to see the curvature of the Earth.

CHAPTER 2. THE NEW GOLDILOCKS

1. Peale, S. J., Cassen, P., & Reynolds, R. T. (1979). Melting of Io by tidal dissipation. *Science*, *203*(4383), 892–894.

2. Cassen, P., Reynolds, R. T., & Peale, S. J. (1979). Is there liquid water on Europa? *Geophysical Research Letters*, *6*(9), 731–734. They also published a correction to their calculations a year later: Cassen, P., Peale, S. J., & Reynolds, R. T. (1980). Tidal dissipation in Europa: A correction. *Geophysical Research Letters*, *7*(11), 987–988.

3. For a summary of the availability of these elements, see Hand, K. P., Chyba, C. F., Priscu, J. C., Carlson, R. W., & Nealson, K. H. (2009). Astrobiology and the potential for life on Europa. In Europa, edited by R. T. Pappalardo, W. B. McKinnon, & K. Khurana. Tucson: University of Arizona Press, pp. 589–629.

CHAPTER 3. THE RAINBOW CONNECTION

1. Jackson, M. W. (2000). Spectrum of Belief: Joseph von Fraunhofer and the craft of precision optics. Cambridge, MA: MIT Press, p. 284.

2. Cruikshank, D. P. (2005). Vassili Ivanovich Moroz: An appreciation. *Lunar and Planetary Science*, *36*, Part 3, conference paper, https://www.lpi.usra.edu/meetings/lpsc2005/pdf/1979.pdf.

3. Some readers may argue that Arthur C. Clarke's novel, and Stanley Kubrick's film, *2001: A Space Odyssey* provided the first imaginative fodder for an ocean within Europa, but they would be wrong. The novel and film were both released in 1968, and in the novel Clarke's moon of choice was actually Saturn's Iapetus. Kubrick, meanwhile, made an inspired choice to base the movie on Europa instead of Iapetus, but the iconic Monolith steals the show and Europa is not described in detail. Kubrick's rationale for choosing Europa was likely to focus on Jupiter instead of Saturn, and Europa ended up being the lucky moon. Fast forward to 1982 and Clarke's *2010: Odyssey Two* changes the scene to Europa, and Kubrick's 1984 adaptation continues with Europa. In the book and movie of *2010*, Europa is featured prominently, but in both cases the plots were heavily informed by the *Voyager* flybys of the late 1970s. By then, scientists were already dreaming of an ocean locked below a shell of ice.

4. In the wake of the *Voyager 2* flyby, Steve Squyres teamed up with Reynolds, Cassen, and Peale to put some of the first, very useful constraints on the nature of Europa's putative subsurface ocean: Squyres, S. W., Reynolds, R. T., Cassen, P. M., & Peale, S. J. (1983). Liquid water and active resurfacing on Europa. *Nature*, *301*(5897), 225.

CHAPTER 5. HOW I LEARNED TO LOVE AIRPORT SECURITY

1. O'Neil, W. J., Ausman, N. E., Gleason, J. A., Landano, M. R., Marr, J. C., Mitchell, R. T., . . . & Smith, M. A. (1997). Project *Galileo* at Jupiter. *Acta astronautica*, *40*(2–8), 477–509.

2. While Io's magnetic field signature was originally, and correctly, attributed to the ionization of material from the volcanoes, Kivelson and her team recently posited that some of the other components of Io's magnetic field signature could be the result of a molten rock mantle that is conductive and can also alter the magnetic field environment. The data fits that model well, and I think they are right!

3. Anderson, J. D., Lau, E. L., Sjogren, W. L., Schubert, G., & Moore, W. B. (1997). Europa's differentiated internal structure: Inferences from two Galileo encounters. *Science*, *276*(5316), 1236–1239.

CHAPTER 7. THE QUEEN OF CARBON

1. Iess, L., Jacobson, R. A., Ducci, M., Stevenson, D. J., Lunine, J. I., Armstrong, J. W., . . . & Tortora, P. (2012). The tides of Titan. *Science*, *337*(6093), 457–459.

2. See, for example, Baland, R. M., Tobie, G., Lefèvre, A., & Van Hoolst, T. (2014). Titan's internal structure inferred from its gravity field, shape, and rotation state. *Icarus*, *237*, 29–41. And Mitri, G., Meriggiola, R., Hayes, A., Lefèvre, A., Tobie, G., Genova, A., Lunine, J. I. and Zebker, H. (2014). Shape, topography, gravity anomalies and tidal deformation of Titan. *Icarus*, *236*, 169–177.

3. Béghin, C., Randriamboarison, O., Hamelin, M., Karkoschka, E., Sotin, C., Whitten, R.C., . . . & Simões, F.(2012). Analytic theory of Titan's Schumann resonance: Constraints on ionospheric conductivity and buried water ocean. *Icarus*, *218*(2), 1028–1042.

4. Note that the Schumann resonance data provide a slightly different, but perhaps better-constrained, estimate for Titan's ice shell thickness compared to the result from the gravity data: 55–80 km versus <100 km.

CHAPTER 8. OCEANS EVERYWHERE

1. Greeley, R., Chyba, C. F., Head, J. W., McCord, T., McKinnon, W. B., Pappalardo, R. T., & Figueredo, P. H. (2004). Geology of Europa. In Jupiter: The Planet, Satellites and Magnetosphere, vol. 2, edited by F. Bagenal, T. E. Dowling, & W. B. McKinnon. Cambridge: Cambridge University Press, 329–362.

2. Zahnle, K., Dones, L., & Levison, H. F. (1998). Cratering rates on the Galilean satellites. Icarus, 136(2), 202–222. See also: Zahnle, K., Schenk, P., Levison, H., & Dones, L. (2003). Cratering rates in the outer solar system. Icarus, 163(2), 263–289.

3. See for example: Vance, S., Bouffard, M., Choukroun, M., & Sotin, C. (2014). Ganymede's internal structure including thermodynamics of magnesium sulfate oceans in contact with ice. Planetary and Space Science, 96, 62–70.

4. Schenk, P. M., Chapman, C. R., Zahnle, K., & Moore, J. M. (2004). Ages and interiors: The cratering record of the Galilean satellites. Jupiter: The Planet, Satellites and Magnetosphere, vol. 2, edited by F. Bagenal, T. E. Dowling, & W. B. McKinnon. Cambridge: Cambridge University Press, 427. See also abstracts from the Lunar & Planetary Science Conferences, with lead authors Gerhard Neukum and Roland Wagner.

5. McCord, T. A., Carlson, R. W., Smythe, W. D., Hansen, G. B., Clark, R. N., Hibbitts, C. A., ... & Johnson, T. V. (1997). Organics and other molecules in the surfaces of Callisto and Ganymede. Science, 278(5336), 271–275.

6. Carlson, R. W. (1999). A tenuous carbon dioxide atmosphere on Jupiter's moon Callisto. Science, 283(5403), 820–821.

7. Soderblom, L. A., Kieffer, S. W., Becker, T. L., Brown, R. H., Cook, A. F., Hansen, C. J., ... & Shoemaker, E. M. (1990). Triton's geyser-like plumes: Discovery and basic characterization. Science, 250(4979), 410–415. And: Kirk, R. L., Brown, R. H., & Soderblom, L. A. (1990). Subsurface energy storage and transport for solar-powered geysers on Triton. Science, 250(4979), 424–429. Brown, R. H., Kirk, R. L., Johnson, T. V., & Soderblom, L. A. (1990). Energy sources for Triton's geyser-like plumes. Science, 250(4979), 431–435.

8. Schenk, P. M., & Zahnle, K. (2007). On the negligible surface age of Triton. Icarus, 192(1), 135–149.

9. Stern, S. A., Bagenal, F., Ennico, K., Gladstone, G. R., Grundy, W. M., McKinnon, W. B., ... & Young, L. A. (2015). The Pluto system: Initial results from its exploration by New Horizons. Science, 350(6258). And Stern, S. A., Grundy, W. M., McKinnon, W. B., Weaver, H. A., & Young, L. A. (2018). The Pluto system after New Horizons. Annual Review of Astronomy and Astrophysics, 56, 357–392.

10. Nimmo, F., Hamilton, D. P., McKinnon, W. B., Schenk, P. M., Binzel, R. P., Bierson, C. J., ... & Olkin, C. B. (2016). Reorientation of Sputnik Planitia implies a subsurface ocean on Pluto. Nature, 540(7631), 94. See also: Kamata, S., Nimmo, F., Sekine, Y., Kuramoto, K., Noguchi, N.,

264 NOTES TO CHAPTER 11

Kimura, J., & Tani, A. (2019). Pluto's ocean is capped and insulated by gas hydrates. *Nature Geoscience*, *12*(6), 407–410.

11. Stevenson, D. J. (1999). Life-sustaining planets in interstellar space? *Nature*, *400*(6739), 32.

CHAPTER 10. ORIGINS IN AN ALIEN OCEAN

1. Note that for this large creature to exist, oxygen must be dissolved in the ocean water. It breathes oxygen in and uses oxygen to eat the microbes it filters through its body. As we see in the next chapter, oxygen is critical to all large life forms on Earth. Europa may actually have a significant amount of oxygen dissolved in its ocean.

2. Note that some methanogens can eat all sorts of small compounds, like vinegar, and generate methane. Hydrogen is essential, but they can use other forms of carbon.

3. Hoehler, T. M., & Jørgensen, B. B. (2013). Microbial life under extreme energy limitation. *Nature Reviews Microbiology*, *11*(2), 83.

4. To be specific, pH is the negative of the logarithm of the proton concentration, as measured in moles per liter. So a pH of 12 is a concentration of 10^{-12} moles of protons per liter of fluid, and a pH of 3 is a concentration of 10^{-3} moles of protons per liter. The lower the pH, the more protons. One mole is 6.022×10^{23} (Avogadro's constant).

CHAPTER 11. BUILDING AN OCEAN WORLD BIOSPHERE

1. Note that a form of photosynthesis that does not produce oxygen (anoxygenic) evolved before oxygen-producing photosynthesis.

2. For a great read on evolutionary innovations, see Falkowski, P. (2015). *Life's Engines*. Princeton: Princeton University Press.

3. These so-called cycloid fractures were very elegantly explained by Dr. Greg Hoppa, Professor Rick Greenberg, and colleagues in Hoppa, G. V., Tufts, B. R., Greenberg, R., & Geissler, P. E. (1999). Formation of cycloidal features on Europa. *Science*, *285*(5435), 1899–1902.

4. This calculation assumes that the heat flux from tidal dissipation is 100 mW/m^2 on Europa's surface. That is a reasonable high-end estimate (Barr, A. C., & Showman, A. P. [2009]. Heat transfer in Europa's icy shell. In *Europa*, edited by R. T. Pappalardo, W. B. McKinnon, & K. Khurana. Tucson: University of Arizona Press, 405–430), but it could be higher (Tobie, G., Choblet, G., & Sotin, C. [2003]. Tidally heated convection: Constraints on Europa's ice shell thickness. *Journal of Geophysical Research: Planets*, *108*[E11]). The low-end limit is about 10 mW/m^2, which is the heating from only the decay of heavy radiogenic elements in the rocks within Europa.

5. Pappalardo, R. T., Head, J. W., Greeley, R., Sullivan, R. J., Pilcher, C., Schubert, G., . . . & Goldsby, D. L. (1998). Geological evidence for solid-state convection in Europa's ice shell. *Nature*, *391*(6665), 365.

6. Hussmann, H., & Spohn, T. (2004). Thermal-orbital evolution of Io and Europa. *Icarus*, *171*(2), 391–410.

7. Phillips, C. B., McEwen, A. S., Hoppa, G. V., Fagents, S. A., Greeley, R., Klemaszewski, J. E., . . . & Breneman, H. H. (2000). The search for current geologic activity on Europa. *Journal of Geophysical Research: Planets*, *105*(E9), 22579–22597.

8. Roth, L., Saur, J., Retherford, K. D., Strobel, D. F., Feldman, P. D., McGrath, M. A., & Nimmo, F. (2014). Transient water vapor at Europa's south pole. *Science, 343*(6167), 171–174. Sparks, W. B., Hand, K. P., McGrath, M. A., Bergeron, E., Cracraft, M., & Deustua, S. E. (2016). Probing for evidence of plumes on Europa with HST/STIS. *The Astrophysical Journal, 829*(2), 121.

CHAPTER 12. THE OCTOPUS AND THE HAMMER

1. Slartibartfast is a character from Douglas Adams's *The Hitchhiker's Guide to the Universe*. This quote is from the third volume, *Life, the Universe, and Everything* (1982). In the books Slartibartfast is a famous planet builder.

2. Morris, S. C. (2003). Life's solution: Inevitable humans in a lonely universe. Cambridge: Cambridge University Press. I highly recommend this book for a detailed examination of evolution and the balance between contingent and convergent solutions.

3. Beatty, J. T., Overmann, J., Lince, M. T., Manske, A. K., Lang, A. S., Blankenship, R. E., . . . & Plumley, F. G. (2005). An obligately photosynthetic bacterial anaerobe from a deep-sea hydrothermal vent. *Proceedings of the National Academy of Sciences, 102*(26), 9306–9310. Note that even though these bacteria can apparently photosynthesize, most of their molecular machinery for photosynthesis likely comes from evolution on the sun-bathed surface of the Earth.

4. Van Dover, C. L., Reynolds, G. T., Chave, A. D., & Tyson, J. A. (1996). Light at deep-sea hydrothermal vents. *Geophysical Research Letters, 23*(16), 2049–2052. See also: Jinks, R. N., Markley, T. L., Taylor, E. E., Perovich, G., Dittel, A. I., Epifanio, C. E., & Cronin, T. W. (2002). Adaptive visual metamorphosis in a deep-sea hydrothermal vent crab. *Nature, 420*(6911), 68.

5. Widder, E. A. (2010). Bioluminescence in the ocean: Origins of biological, chemical, and ecological diversity. *Science, 328*(5979), 704–708.

6. Goodman, J. C., Collins, G. C., Marshall, J., & Pierrehumbert, R. T. (2004). Hydrothermal plume dynamics on Europa: Implications for chaos formation. *Journal of Geophysical Research: Planets, 109*(E3).

7. For more information on these fascinating creatures, see the website of Cornell University professor John P. Sullivan, arguably the world's expert on Mormyridae: http://www.mormyrids.myspecies.info.

8. Simon Conway Morris' book, cited earlier, has a significant section dedicated to the mormyrids, and to the South American gymnotids, which independently evolved electroreception capabilities similar to those of the mormyrids.

9. By non-biological I simply mean that plenty of "tools" emerge from evolution alone, but they are often things like antlers, horns, claws, etc., that serve a purpose in hunting and defense. Also, one could certainly argue that ants, bees, termites and other insects are "farmers" (think aphids), and I see no reason why, on some worlds, insects could not become sentient creatures, in part due to the selection pressure of tool use.

10. To be more specific, the vent fluid is likely feeding microbes that are symbiotically packaged into a tiny backpack-like organ on the shrimp. The bacteria eat compounds in the vent fluid, and they generate organics that the shrimp can eat. The bacteria also detoxify the vent water, transforming various metal and sulfur compounds into benign materials.

11. For more on this incredible ecology, see the now-classic text, Van Dover, C. (2000). The ecology of deep-sea hydrothermal vents. Princeton University Press.

12. Früh-Green, G. L., Kelley, D. S., Bernasconi, S. M., Karson, J. A., Ludwig, K. A., Butterfield, D. A., Boschi, C., & Proskurowski, G. (2003). 30,000 years of hydrothermal activity at the Lost City vent field. *Science, 301*(5632), 495–498.

13. Benyus, J. M. (1997). Biomimicry: Innovation inspired by nature. New York: William Morrow.

14. Two of my favorite examples of minerals made by microbes are magnetite made by magnetotactic bacteria and silicon shells made by diatoms. Magnetotactic bacteria grow chains of nanometer-sized crystals of the magnetic mineral magnetite (Fe_3O_4). The chain of magnetite serves as a tiny compass, giving the microbe the ability to line up with the Earth's magnetic field. Within oceans, lakes, or sediments, the Earth's magnetic field runs nearly vertical, and the microbe spins its tail (flagellum) to move up or down along the direction of the magnetic field, searching for the chemical sweet spot (e.g., just the right amount of dissolved oxygen in the water). By aligning itself with the magnetic field and moving up and down, the microbe can reduce its search region from three dimensions to just one dimension—a thin, horizontal layer of water or sediments. The other example of minerals made by microbes is the silicon shells of diatoms, which are intricate and stunning bulbs of glass, i.e., silica, that house the organism within. The silicon atoms the microbe uses to make its shell come from ocean water and can be sourced from geologically active regions, like hydrothermal vents.

CHAPTER 13. A PERIODIC TABLE FOR LIFE

1. Benner, S. A., Ricardo, A., & Carrigan, M. A. (2004). Is there a common chemical model for life in the universe? *Current Opinion in Chemical Biology, 8*(6), 672–689.

2. Here I use the term "basic" as in acids and bases.

3. Benner, Ricard, & Carrigan, Is there a common chemical?

4. Curiously, minerals (like life) lack a universally accepted definition. All definitions are some variation of what I have provided here, which is from: Klein, C. (2002). *Manual of Mineral Science*, 22nd ed. Hoboken, NJ: John Wiley and Sons. One example of a debatable point in this particular definition is the term "naturally occurring." We can now make diamonds and other inorganic crystals through industrial processes; therefore, since they did not occur naturally, are they not minerals?

5. See the above cited article Benner et al. (2004) for more detail on polysilanes.

6. Bains, W. (2004). Many chemistries could be used to build living systems. *Astrobiology, 4*(2), 137–167.

7. Hoshika, S., Leal, N. A., Kim, M. J., Kim, M. S., Karalkar, N. B., Kim, H. J., . . . & Benner, S. A. (2019) Hachimoji DNA and RNA: A genetic system with eight building blocks. *Science, 363*(6429), 884–887.

8. Lepper, C. P. (2015). Effects of high pressure on DNA and its components. Doctoral dissertation, Massey University, Manawatū, New Zealand.

9. These polymers (polysaccharides) are used for structures in the cell wall, and they are also stored for energy reserves (e.g., carbohydrates). The ability of sugars to polymerize is also

central to the structure of DNA and RNA. "Ribo" in ribonucleic acid refers the ribose sugar backbone that links the nucleic acids.

10. Westheimer, F. H. (1987). Why nature chose phosphates. *Science, 235*(4793), 1173–1178.

CHAPTER 14. SEEKING SIGNS OF LIFE

1. Woese, C. R., & Fox, G. E. (1977). Phylogenetic structure of the prokaryotic domain: The primary kingdoms. *Proceedings of the National Academy of Sciences, 74*(11), 5088–5090.

2. The term "organics" is typically used for carbon compounds that have two or more carbons linked together. CO_2, for example, is not a carbon compound. Some researchers also argue that methane, which has only one carbon atom linked to four hydrogens, is also not an organic molecule. Importantly, graphite, which consists of nothing but carbon linked to carbon, is also not an organic compound.

3. Ming, D. W., Archer, P. D., Glavin, D. P., Eigenbrode, J. L., Franz, H. B., Sutter, B., . . . & Mahaffy, P. R. (2014). Volatile and organic compositions of sedimentary rocks in Yellowknife Bay, Gale Crater, Mars. *Science, 343*(6169), 1245267.

4. Klein, H. P. (1998). The search for life on Mars: What we learned from *Viking*. *Journal of Geophysical Research: Planets, 103*(E12), 28463–28466. Chyba, C. F., & Phillips, C. B. (2001). Possible ecosystems and the search for life on Europa. *Proceedings of the National Academy of Sciences, 98*(3), 801–804.

5. Lovelock, J. E. (1965). A physical basis for life detection experiments. *Nature, 207*(997), 568–570.

6. The handedness of molecules comes from how the molecule affects the rotation of plane-polarized light.

7. Sagan C., Thompson, W. R., Carlson, R., Gurnett, D., & Hord, C. (1993) A Search for life on Earth from the *Galileo* spacecraft. *Nature, 365*, 715–721.

CHAPTER 15. A NEW AGE OF OCEAN EXPLORATION

1. See for example, Hand, K. P., and German, C. R. (2017). Exploring ocean worlds on Earth and beyond. *Nature Geoscience, 11*, 2–4.

2. A brief note on budgets and money that is used for innovating in ocean exploration and in space exploration. NASA has a much larger budget than the combined budgets of NSF and NOAA. Up until recently, however, there were not commercial incentives to explore space—companies working in the space sector were on contracts with NASA or the military. Ocean exploration, meanwhile, has a very strong commercial sector. From shipping, to fishing, to oil and gas exploration, there are many ways companies invest in the exploration (or exploitation, one might argue) of our ocean. The oil and gas industry, for example, has no interest in going to the Moon, but they spend billions of dollars every year in the Gulf of Mexico. If the Moon had oil, that might be a different story. Federal funds are typically used to explore new realms to catalyze new commercial opportunities (this is actually in NASA's charter). In recent years, the commercial sector in space has started to expand rapidly, thanks in large part to the desire for imagery of the Earth from space and improved transmission of data around the globe. Smartphones are

everywhere, and so are the satellites that feed them GPS coordinates, updated maps, and the occasional text from around the globe. Nevertheless, the total investment from industry and government funds is much larger in the realm of our ocean than it is for space.

3. It is a little difficult to know whom to cite for this phrase. The astronaut Don Pettit wrote a great essay with this title in 2012 (https://www.nasa.gov/mission_pages/station/expeditions /expedition30/tryanny.html), but I am almost certain this phrase became popular well before then. Ultimately, the rocket equation itself is attributable to the legendary Russian scientist Konstantin Tsiolkovsky, who published the equation in 1903 (although at least two mathematicians derived the equations prior to Tsiolkovsky).

4. NASA. (2011). Mars Science Laboratory Launch. https://www.jpl.nasa.gov/news/press _kits/MSLLaunch.pdf.

5. Note that for Mars we are now at a point where returning samples back to Earth is the top priority. The many landed vehicles that have explored Mars have made it worthwhile, according to the Mars community, to invest in bringing samples back. This is an exciting and challenging mission that may help us find signs of ancient life in the rocks from Mars.

6. The *Galileo* and *Cassini* spacecraft were not extensively sterilized before launch, and thus there was a small chance that earthly microbes might have hitched a ride. If either spacecraft died while in orbit, it might have eventually crashed into Europa or Enceladus, increasing the risk of contamination. For this reason, while there was still enough fuel in their engines, the spacecraft were sent into their respective planets to be burned up.

7. I often get asked about private funding for such missions. Unfortunately, for the time being, these missions are too expensive for a private donor to fund and build. At a few billion dollars per mission, there's a limited cadre of viable donors. That is not to say it is not possible. Some individuals could make it happen. And, who knows, maybe a giant crowdfunding project could work.

8. National Research Council. (2003). New frontiers in the solar system: An integrated exploration strategy. Washington, DC: The National Academies Press. https://doi.org/10.17226 /10432.

9. National Research Council. (2011). Vision and voyages for planetary science in the decade 2013–2022. Washington, DC: The National Academies Press. https://doi.org/10.17226/13117.

10. Former congressman John Culberson worked tirelessly to increase NASA's science budget for Earth sciences and space sciences. He was, and continues to be, a strong champion for the exploration of Europa and ocean worlds. Beginning in the mid-2000s up to 2018, he worked to fund the *Clipper* mission and the lander mission. He also ensured that NASA's full mission and research portfolio was well-funded. When he lost the election in 2018, much of that funding became uncertain. President Trump provided no funding for a possible Europa lander in his budget for fiscal year 2020. With regard to the planetary science community and NASA, see the following report for an assessment of why progress on the lander mission concept was halted: Office of Inspector General. (2019). Management of NASA's Europa mission. https://oig.nasa .gov/docs/IG-19-019.pdf.

11. Roosevelt, T. (1899). The strenuous life. Speech given in Chicago on April 10. https:// voicesofdemocracy.umd.edu/roosevelt-strenuous-life-1899-speech-text/.

12. NASA. (2016). Europa lander study report. https://europa.nasa.gov/resources/58/europa -lander-study-2016-report/.

13. As an important aside, one of the most exciting aspects of working on the design for the Europa lander was the chance to work with many great scientists and engineers. I get to come up with crazy science questions, and then the engineers are tasked with figuring out how to get it done. They are the truly brilliant ones who solve many problems behind the scenes. One of the recurring themes I hear from the engineers has been that many of these missions can be done without new technology. They are hard and challenging missions, but they require no magic wands, no need to bend the laws of physics.

14. Galilei, G. (2016). Sidereus nuncius, or The sidereal messenger. Chicago: University of Chicago Press. Originally published in Venice, Italy, in 1610.

INDEX

Note: page numbers followed by "f" and "n" indicate figures and endnotes, respectively.

accretional heat, 122

acetylene, 210

Adams, Douglas, 184

adenosine diphosphate (ADP), 222

adenosine triphosphate (ATP), 14, 222–23

airport security analogy, 79–80, 90, 91. *See also* magnetic fields and magnetometer data

alternative biochemistries, 18

Alvin, 22–24

ammonia: snow line for, 41f, 44; as solvent, 209–10, 211

Anderson, John, 74–75

anoxygenic photosynthesis, 264n1(ch11)

"architectural blueprint," 142

argon, 111–12, 119

Ariel, 42f

Arnold, Frances, 216

Asimov, Isaac, 226

Atlantic Ocean. *See* oceans of Earth

Atlantis Space Shuttle, 71

atmospheres: Callisto, 127; Earth, 162–63, 243–44; Pluto, 131; Titan, 109–12, 119; Triton, 128

atomic mass units (amu), 99–100

autonomous underwater vehicles (AUVs), 255

bacteria: cyanobacteria, 161–62; infrared photon photosynthesis and, 187; magnetotactic bacteria, 266n14; mitochondria and, 163

Barton, Otis, 20–21

bathyscaphes, 21, 261n6

bathyspheres, 20

battery analogy, 144–46, 222–24

Beebe, William, 20–21

Béghin, Christian, 118

Benner, Steven, 211–12, 218–19, 221

Bierhaus, Beau, 170

big ship, big science approach, 247–48

binary nucleotide system, 217–18

biochemistries, alternative, 18

biodomes scenario, 204–5

biology, revolutionizing of, 207–9

bioluminescence, 8, 187–88

biomimicry, 201

biosignatures and detection of life: about, 229–30; compositional, morphological, and context evidence, 230, 237–42; dual-use measurements, 236; future landers on Enceladus and Europa, 238, 242–43; *Galileo* collection of Earth evidence, 243–44; instruments, selection of, 230–31; Mars *Viking* landers experiments, 231–37; robust, redundant, and complementary measurements, 236–37, 242

biosphere-building and Europa: Gaia hypothesis and, 181–82; geologic cycling of ice, 178–81; grain size of ice and, 182; ice shell thickness question, 168–78, 180; irradiated ice, chemistry of, 165–68; oxidant availability issue, 160–65

bonds, covalent, 29, 30f, 212–13

bottom-up approach to origin of life, 143, 146, 147–48

Bronze Age, 199

Brown, Mike, 67

Bunsen, Robert, 56–58

Bunsen burner, 56

Cairns-Smith, Graham, 158

Callisto: evidence for ocean, 125–28; gravity data, 127; gravity well of, 71f, 72; ice-covered oceans on, 10, 13; ice III on seafloor, 127; ice surface, discovery of, 60; induced magnetic field, 125, 127; Laplace resonance and, 38; nitrogen, 44; size and density, 42f, 44–45, 125; surface age, composition, and craters, 125–27, 126f; tidal heating, 37; *Voyager* images of, 60–61

Calvin, Wendy, 167–68

Cambrian explosion, 163–64

Cameron, James, 4, 7, 11, 152–53

carbon: as basic building block, 212–13; carbon-13 (^{13}C), 241–42; carbon-14 (^{14}C), 233, 241; carbon–carbon bonds, 215; carbon–silicon bonds, 216; conversion into different phases, 214; isotope fractionation and measurement, 241–42; polarity and, 210; solvents and, 212

carbon dioxide: Callisto, 127; Enceladus, 100; as ice on Europa, 43–44; Mars *Viking* landers experiments, 233–34; methanogenic microbes and, 156–57; as oxidant, 146; photosynthesis and, 161–62; snow line for, 40, 41f

Carlson, Bob, 62, 64–65, 67, 167, 243

Carr, Mike, 171

Casani, John, 87

Cassen, Patrick, 36, 61, 262n4(ch4)

Cassini: death of, 251, 268n6; Enceladus plume sampling, 98–104; Ganymede flyby, 121; *Huygens* probe (Titan), 110–12, 116, 118; mission of, 97; Titan flybys, 113–15; water likelihood and, 13

catalysis, 140, 158

Caterson, Tym, 152–53

Ceres, 13

Challenger expedition to Mariana Trench, 19, 247

Charon, 42f, 132

chemical disequilibrium, 144, 146–47, 157, 243–44

chemistry–spectroscopy link, 56–59

chemosynthesis, 23, 154

chirality, 240–41

CHNOPS elements (carbon, hydrogen, nitrogen, oxygen, phosphorus, and sulfur), 39–44, 222. *See also specific elements*

Chyba, Chris, 39–40, 140, 143

citric acid, 223

Clarke, Arthur C., 225–26, 262n3(ch4)

clathrates, 133

Clipper, 245–46, 252–53, 268n10

clocks, synchronized, 74

cnidarian, 153

coin sorter analogy, 98–99. *See also* mass spectrometers

comets, 100–101

compartmentalization, 141

complementary biosignatures, 236–37, 242

compositional biosignatures, 230, 238–42

computation, 202, 205

context biosignatures, 230, 237–38

contingent vs. convergent adaptations, 185

convection, 173–74

cooking, 198

Copernican Revolution, 15, 257, 259

Cosmic Dust Analyzer (CDA), 98, 101–2

Cousteau, Jacques-Yves, 18

covalent bonding, 29, 30f, 212–13

Crane, Jocelyn, 20

craters: Callisto, 126, 126f; Enceladus, 96–97, 170; Europa, 61, 170; Pluto, 131–32

crown glass, 52

cryovolcanism, 110

Culberson, John, 268n10

Curiosity rover, 14, 235, 249–50

cyanobacteria, 161–62

Dalton, Brad, 67

Dalton, John, 56

"Dare Mighty Things" motto, 253–54

Deamer, David, 141

Deep Rover submersibles, 152–53

Deep Space Network (DSN), 73–74, 76, 96

dehydration, 147

density: change in water across the liquid–solid boundary, 28; gravity wells and, 73; of planets and moons, 41–44, 42f. *See also specific worlds*

detection of life. *See* biosignatures and detection of life

diapirs, 174, 176, 178

diatoms, 215, 266n14

Dione, 42f

dipole monomers, 221

diving bells, 20

DNA (deoxyribonucleic acid): alternative biochemistries, 14, 18, 216–19; as blueprint molecule, 142, 216–17; breakdown of, 14; discovery of structure of, 232; quaternary, binary, and octal systems, 217–19

Doppler shift data, 74, 116

Dragonfly mission, 256–57

dual-use measurements, 236

Earth: atmosphere of, 162–63, 243–44; CHNOPS elements and, 40; distance from sun, 25; *Galileo* data on, 243–44; life potentially spreading from, 16–18; photosynthesis transition, 161–63; tides and tidal bulge, 33–35, 35f; time for origin of life on, 139–40. *See also* oceans of Earth

echolocation, 190–91

Einstein, Albert, 70

Elachi, Charles, 253

electricity, 200

electroreception, 191–92

elliptical orbits and tidal heating, 35, 37–38

Enceladus: age of, 106–7; *Cassini* mission and photos of plumes, 97–98; chemistry to support microbes, 103–4; craters, 96–97, 170; elliptical orbit, 35; future landing missions, 238, 250, 255; geologic activity on, 96; "glorified comet" hypothesis, 100–101; hydrothermal plumes, 189; hydrothermal vents scenario and, 150; ice-covered oceans on, 10, 13; irradiation of ice, 167; mass spectrometers and plume analysis, 98–104, 159; in new Goldilocks zone, 45; nitrogen and carbon, 44; return of sample to Earth (potential), 236; serpentinization on, 159; size and density, 42f, 43, 45, 96; tidal heating, 35; water–rock interaction, 43; wobbles and size of ocean on, 104–6

Epsom salts, 66–67

erasable-programmable read-only memory (EPROM), 202

ethane, liquid, 109, 113, 115

ethylene, 210, 220–21

Europa: in *2001: A Space Odyssey*, 262n3(ch4); CHNOPS elements, 43–44; future landing missions, 242–43, 245, 250, 254, 258; geologic cycling of ice, 178–81; gravity signature, 69–78; gravity well of, 71f, 72–73; heat conduction and flow, 172–77; hydrothermal vents scenario and, 150; ice grain size, 182; ice shell, 60, 61, 63f; ice shell thickness, 168–78, 180, 188–89; induced magnetic field and salty ocean hypothesis, 89f, 91–94; Laplace resonance and, 37–38; liquid water ocean, summary of evidence for, 94–95; moment of inertia, 75–77, 78f; NASA mission studies, 245; in new Goldilocks zone, 37, 45; non-ice material (salts or sulfuric acid), 65–68, 179; oblateness of, 72; oceans, ice-covered, 10, 13; orbit, 35, 37–38, 175; oxygen, dissolved, 264n1(ch10); plumes, 176, 189; rocks from Earth, simulation of, 18; size, density, and mass, 42f, 43, 45, 72–73; snow, 177; spectroscopy, 60–68, 60f; sulfonamide and, 221; surface age, 169–70; surface craters, 61, 170; surface fractures, chaos regions, and freckles

Europa (cont.)
  (lenticulae), 170–72, 171f, 172f, 174–78, 175f; temperature of surface and ocean, 173; three-layer model of, 77–78, 78f; tidal heating, 35, 36, 174–76, 264n4(ch11)
Europa *Clipper*, 245–46, 252–53, 268n10
evidence for life. *See* biosignatures and detection of life
evolution: Cambrian explosion, 163–64; civilization, 195–200; contingent vs. convergent adaptations, 185; photosynthesis transition, 161–63; sensory systems, 186–92; technomimicry, 201–5; tools and technology, 192–95. *See also* biosphere-building and Europa
exothermic reactions, 154–55, 197
extracellular polymeric substances (EPS), 182
eyes, evolution of, 186–89

Farley, Tom, 81
farming, 194, 195–98
fire as tool, 194–95, 198
flash memory, 202–4
flint glass, 52
floating-gate transistors, 203–4
food chains, photosynthesis- vs. chemosynthesis based, 23, 154
foraminifera, 229
Fox, George, 232
Fraunhofer, Joseph Ritter von, 51–58
"freckles" (lenticulae), 174–76, 175f
frost lines. *See* snow lines

Gaia hypothesis, 164, 181–82, 205
Galápagos Rift, 22–24
Galileo: carbon dioxide ice found on Europa, 43–44; death of, 251, 268n6; Earth data gathered by, 243–44; Europa images, 65–66, 171; Ganymede flybys, 121; gravity measurements of Europa, 69–78; high-gain antenna failure and data issues, 83–86; launch and slingshots, 70–71;

magnetometer and Jupiter system magnetic fields, 79, 82–83, 86–94; NIMS spectrometry, 62–67; photopolarimeter, 173; radiation chemistry data, 167; tracking with DNS, 73–75; water likelihood and, 13
Galileo Galilei, 15, 257, 259
Ganymede: evidence for ocean, 121–25; furrow systems, 122; future missions, 252; gravity well of, 71f, 72; ice-covered oceans on, 10, 13; ice III on seafloor, 124; ice surface, discovery of, 60; Laplace resonance and, 37–38; magnetic field, 90–91, 121–22; in new Goldilocks zone, 37; orbit, 35, 37–38, 175; oxygen and ozone in ice of, 168; size and density, 42f, 44–45, 124; surface composition, 122–23, 123f; tidal heating, 35, 90, 122; *Voyager* images of, 61
Gas Chromatograph–Mass Spectrometer (GCMS), 233–35
Gas Exchange experiment (*Viking* landers), 232
geometric librations, 104–5
German, Chris, 189
germanium, 215
Gladman, Brett, 17–18
glass, refractive properties of, 52–54
Goldilocks model, new: edge of, 125; geochemical factors and size/density of worlds, 39–45; ice, floating and crystal structure of, 25–32; old Goldilocks model, 25–26; tidal heating in moons, 32–38
gravity and gravity signatures: Callisto, 71f, 72, 127; Europa, 69–78, 71f, 78f; Ganymede, 71f, 72; gravity wells, 70–74, 71f; Io, 71f, 72; measured by tracking a spacecraft, 73; moment of inertia, 75–78, 78f; in ocean exploration, 6; shape of space and, 69–70, 71f; tidal heating and gravitational force, 32–33; Titan k_2 Love number and gravity signature, 116–17

"Great Tree of Life," 224–26, 225f

Greely, Ron, 245

Greenberg, Rick, 171–72

greenhouse effect, solid-state, 129

habitability: Gaia hypothesis and, 164; habitable zone, 26; "inhabited" vs., 137; keystones for, 139. *See also* Goldilocks model, new

habitability, evidence for. *See* gravity and gravity signatures; oceans of the outer solar system; spectroscopy

hachimoji DNA, 219

Halley, Edmund, 20

handedness, 240–41

Hanner, Margaret, 83

hearing, sense of, 190–91

heat conduction on Europa, 172–77

heat sources. *See* hydrothermal vents; radiogenic heating; tidal heating

Herschel, John Frederick William, 58–59

hexagonal crystal lattice of ice, 30f, 31, 32f

Hitchhiker's Guide to the Galaxy (Adams), 184

Hollister, Gloria, 20

Hoppa, Greg, 171–72

Hubble Space Telescope, 67–68, 168, 176

Hussman, Hauke, 175

Huygens probe, 110–12, 116, 118

hydrogen: Enceladus, 102–3; hydrogen atoms in water, 29; hydrogen bonds, 29; liquid, 211–12; methanogenic microbes and, 156–57; as reductant, 146; serpentinizing vents and, 156

hydrogen peroxide, 167–68

hydrogen sulfide, 146

hydrothermal vents: biodomes scenario, 204–5; "blow torch" vs. "hand warmer," 154–56; as chemical oases, 4; discovery of, 22–23; on Enceladus, 102–3; farming, possibility of, 195–98; locating with senses, 189–90; Lost City, 151–59, 197; Menez Gwen, 7–9; methanogenic

microbes and, 156–57, 160, 264n2(ch10); origin-of-life scenario, 138, 149–50; serpentinizing, 154–59, 197; shrimp swarms and sculpting of, 196–97; Snake Pit, 196, 197; stability of, 197

hydroxyl radicals, 167

Iapetus, 42f, 262n3(ch4)

ice (water ice): atomic structure and covalent bonding of water molecules, 30f; crystal structure and floating of, 25–32; dense ice under high pressure, 44–45; freezing point suppressed by salts, 173; geological cycling of, 178–81; grain size of, 182, 205; ice III, V, and VI, 124, 127; infrared spectroscopy of, 59–60; as insulating layer, 27–28, 31; radiation chemistry, 166–68; snow as insulator, 177. *See also under* Europa; Ganymede; Titan

Iess, Luciano, 117

inertia, moment of, 75–77, 78f

information coding and molecules, 141–43, 216–19

infrared photons, 187

infrared spectroscopy, 58–60

intelligence and technology, evolution of: computation, 202, 205; contingent vs. convergent adaptations, 185; cooking, 198; Earth's oceans, tool use in, 192–94; electricity, 200; farming, 194, 195–98; fire, 194–95, 198; Gaia hypothesis and, 205; memory, non-volatile, 202–4; metal-working, 199–200; mobility, 193–94, 195; night sky and sense of wonder, 184, 205–6; senses, 186–92; technomimicry, 201–5

Io: gravity well of, 71f, 72; Laplace resonance and, 37–38; magnetic field, 88–90, 262n2(ch5); orbit, 37–38, 175; tidal heating, 36–37; volcanic activity, 36–37, 61, 88–90; *Voyager* images of, 61

Ion and Neutral Mass Spectrometer (INMS), 98–104

Iron Age, 199

iron as oxidant, 146

Jet Propulsion Laboratory (JPL), 62, 67, 74, 98, 124, 157, 176, 181, 245, 253–54

Joyce, Gerald, 142

Juno, 13, 38

Jupiter and the Jovian system: *Cassini* mission, 97; gravity wells, 71f, 72–73; Laplace resonance and, 37–38; magnetic field, 87–91, 89f, 165–66; moons orbiting, 13, 15, 257, 258f; orbital dance, 175; ten-hour axial spin, 88; tidal heating, 35–38. *See also* Callisto; Europa; Ganymede

Jupiter Icy Moons Explorer (JUICE) mission, 125, 252–53

k_2 Love number and gravity signature, 116–17

Kattenhorn, Simon, 179

Keck Observatory telescopes, 67

Keldysh (Russian research vessel), 5–11

Kelley, Deborah (Deb), 154

Kempf, Sascha, 102

Kepler, Johannes, 33, 104

Khurana, Krishan, 92–93

Kirchhoff, Gustav, 56–58

Kivelson, Margaret, 80–83, 86–87, 92–94, 262n2(ch5)

Kubrick, Stanley, 262n3(ch4)

Kuiper, Gerhard, 60

Kuiper belt, 60, 106, 129

Labeled Release experiment (*Viking* landers), 233

Laplace, Pierre-Simon, 38

Laplace resonance, 37–38, 122, 175

lead, 215

lenticulae ("freckles"), 174–76, 175f

Libby, Willard, 81

librations, physical and optical, 104–6

life: alternative biochemistries, 18; biology revolutionized by discovery of, 207–9; CHNOPS elements and, 39–44; extant vs. extinct, 14; Gaia hypothesis, 164, 181–82, 205; as "hypothesis of last resort," 244; "impact frustration of," 140; second origin, search for, 16–18; time needed to originate, 106–7; "weird," 18, 113–15, 119, 210. *See also* biosignatures and detection of life; Goldilocks model, new; origin of life

life, key parameters of: battery of biology (ATP and alternatives), 222–24; bricks and mortar (monomers and polymers), 219–22; building-block elements (carbon and alternatives), 212–16; Great Tree of Life or Periodic Table of Life, 224–26, 225f, 226f; information coding (DNA-RNA and alternatives), 216–19; solvents (water and alternatives), 209–12

light spectrum. *See* spectroscopy

limestone, 229

Lorenz, Ralph, 256

Lost City hydrothermal vents, 151–59, 197

Lovelock, James, 164, 181, 239

lux veritatis (the light of truth), 51

magnesium sulfate, 66–67

magnetic fields and magnetometer data: airport security analogy, 79–80, 90, 91; Earth, 165; Europa's induced field and salty ocean hypothesis, 91–94; Ganymede, 90–91, 121–22, 124; induced magnetic field, defined, 80; Jupiter system, 87–91, 89f, 165–66; Kivelson and the *Galileo* magnetometer, 79, 82–87; polarity changes with induced fields, 93–94; Saturn, Enceladus, and, 97; Sun, 165; Titan, 117–18

magnetite, 266n14

magneto-hydrodynamic dynamos (MHDs), 121–22

magnetotactic bacteria, 266n14

manganese oxide, 146
Margulis, Lynn, 209
Mariana Trench, 19, 21–22
Mariner 2, 12, 21
Mars: Callisto compared to, 37; *Curiosity* rover, 14, 235, 249–50; dry and rocky, 40; Ganymede compared to, 90; Goldilocks model and, 25–26; habitability, past possibility of, 14; origin of life conditions and, 149–50; potential life coming from Earth, 16–17; returning samples to Earth, 268n5; robotic exploration of, 208; *Viking* landers mission, 231–37
mass spectrometers, 98–104
McAffrey, Paul, 152–53
Means, Joe, 86
memory, non-volatile, 202–4
Mendeleev, Dmitri, 57, 208
Menez Gwen seamount, 7–9
Mercury: atmosphere, lack of, 109; Ganymede compared to, 45, 90, 121; robotic exploration of, 208
metabolism, 144–47, 157
metalworking, 199–200
methane: Enceladus, 100, 103; methanogenic microbes and, 156–57; with oxygen in Earth's atmosphere, 244; as reductant, 146; snow line for, 40, 41f, 44; Titan, methane meteorological cycle on, 108–9, 113; in Titan atmosphere, 109–10; Triton, 130–31
methanogenic microbes, 156–57, 160, 264n2(ch10)
microbes: cyanobacteria, 161–62; Earth microbes on spacecraft, 250–51; Enceladus chemistry and, 103–4; Europa and possibility of, 180; flash memory and, 204; hydrogen from hydrothermal vents and, 103; metabolism and, 146; organelles, 163
Miller, Stanley, 143
Mimas, 42f, 43
minerals: catalytic properties of, 139, 140, 147; definitions and classifications of,

213–14, 266n4; in early Earth ocean, 162; floating-gate transistors and, 204; fossils, 14; hydrothermal vent theory and, 157–58; as oxidants and reductants, 146; serpentinization and, 154–55; silicate mineralogy, 213, 214; Snake Pit hydrothermal vents and, 196
Miranda, 13, 42f, 120
Mir submersibles, 3–11, 152
mitochondria, 163
mobility, tool use and, 193–94, 195
moment of inertia, 75–77, 78f
monomers, 219–22, 239–41
Moon (of Earth): elliptical orbit and wobble, 104–5; robotic exploration of, 208; size and density, 42f; tides and tidal bulge, 33–35, 35f
moons with ice-covered oceans. *See* oceans of the outer solar system; *specific moons by name*
Moore, Jeff, 170
Morabito, Linda, 61
Mormyridae, 191
Moroz, Vassili, 60
morphological biosignatures, 230
morphological evidence for life, 238
Morris, Simon Conway, 191
multicelled organisms, evolution of, 163
mythology, 205–6

NASA: funding and budgets, 62, 235, 251–52, 267n2(ch15), 268n10; mission studies and proposals to, 82–83, 245, 256; search for life on Mars (1976), 231–36; skilled labor bottleneck, 253. *See also specific missions and locations, such as* Cassini *or* Europa
National Academies of Sciences, 253
National Oceanographic and Atmospheric Administration (NOAA), 247, 267n2(ch15)
National Science Foundation (NSF), 247, 267n2(ch15)

Nealson, Ken, 5
Near-Infrared Mapping Spectrometer (NIMS), 62–67
Neptune, 257. *See also* Triton
New Horizons, 13
Newton, Isaac, 52–53
night sky, 184, 205–6
Nimmo, Francis, 132
nitrate, 146
nitrogen: Europa, 44; liquid, 215; Titan, 109–10; Triton, 128, 130–31
Nixon, Richard, 235
noble gases, 111–12
non-specific organics, 239
Nonsuch Island, 20
non-volatile memory, 202–4
Northern/Southern lights, 165

Oberon, 42f
oceans of Earth: Atlantic *Mir* submersible expedition, 3–11; big ship, big science approach, 247–48; commercial sector, 267n2(ch15); *Deep Rover* submersibles, 152–53; history of exploration of, 19–24; Lost City hydrothermal vents, 151–59, 197; Mariana Trench, 19, 21–22; Snake Pit hydrothermal vents, 196, 197; testing technology in, 246; unmapped, 246–47
oceans of the outer solar system: Callisto, 125–28; dense ice on seafloors, 44–45; Enceladus, 10, 13, 104–6; Europa, 10, 13, 91–95; Ganymede, 121–25; moons with, 10; Pluto, 131–33; rogue planets, 133–34; Titan, 10, 13; total volume of liquid water, 120; Triton, 128–31
octal DNA system, 218–19
octopi, 183, 187, 192, 197–98, 200, 204
Opportunity, 14
Optical Institute, Bavaria, 51–52
optical librations, 104–5
orbits, elliptical. *See* elliptical orbits and tidal heating
organelles, 163

organics (carbon compounds): as biosignatures, 239, 241; Callisto, 127; carbon dioxide and, 44, 234; combined with oxygen, 163; of comets, 101; defined, 267n2(ch14); Enceladus, 100, 103; Mars non-detection of, 234; as reductants, 146; redundant measurements, 237; small, 100; Triton, 129; *Viking* landers experiments, Mars, 232–33
Orgel, Leslie, 142
origin of life: catalytic surfaces, 140; compartmentalization, 141; discovery of life as contingent on, 229; "habitable" vs. "inhabited" and, 137; information storage and replication, 141–43; metabolism, 144–47; primordial soup theory, 143; serpentinizing hydrothermal vents theory, 157–59; time parameter, 139–40; top-down vs. bottom-up approach, 143, 146, 147–48; warm pond/tidepool scenario vs. hydrothermal vent scenario, 138, 148–50
outgassing, 65
oxidants: availability of, 160–65; in Earth atmosphere, 244; ice cycling on Europa and, 179–80; in metabolism, 145–46; serpentinizing vents and, 155
oxygen: biodomes scenario and, 205; Cambrian explosion, 163–64; as critical to life, 264n1(ch10); excess, problem of, 188; on Ganymede, 123, 168; with methane in Earth's atmosphere, 244; as oxidant, 146; photosynthesis and, 162; radiation chemistry and, 167–68; in water, 28
ozone, 123, 167–68

Pappalardo, Bob, 174, 245
Peale, Stanton, 35–36, 61, 262n4(ch4)
peridotite, 155
periodic table: carbon group, 213–14; development of, 56, 57, 208; life and elements of, 39f, 208
"Periodic Table of Life," 224–26, 225f, 226f

peroxide, 146

Pettit, Don, 268n3

pH, 158–59, 211, 264n4(ch10)

Phillips, Cynthia, 176

phosphate, 223–24

phosphonamide, 221

phosphorous, 222–23

photic zone, 7

photopolarimeter, 173

photosynthesis: anoxygenic, 264n1(ch11); food chain and, 23; infrared photons and, 187; oxidant availability and, 161–64; red edge on Earth, 243

physical librations, 104–6

Piccard, Auguste, 21, 261n6

Piccard, Jacques, 21–22

Pioneer, 13, 81–82

plastics, 220–21

Pluto: atmosphere, 131; evidence for ocean, 131–33; oceans, possibility of, 13; radiogenic heating and clathrates, 133; size and density, 42f; Sputnik Planitia, 131–32; surface of, 131–32

PNA (polynucleic acid), 142

polarity, solvents and, 115, 210–12

polymerase chain reaction (PCR), 232

polymers, 219–22, 240, 266n9

polysilanes, 215

Porco, Carolyn, 98

Postberg, Frank, 102

potassium-40, 112

primordial soup theory, 143

private funding, 268n7

Prockter, Louise, 179, 245

proteins, 142, 147, 210, 217, 220–21

protons, pH and, 158, 264n4(ch10)

Pyrolytic Release experiment (*Viking* landers), 233

quaternary nucleotide system, 217–18

racemization, 241

radiation chemistry, 165–68

radiogenic heating: Europa, 264n4(ch11); Pluto, 133; Triton, 130

rainbow connection. *See* spectroscopy

red edge, 243

reductants, 145–46, 244

redundant biosignatures, 236–37, 242

refraction index, 52, 54

reproduction, 141–42

Reynolds, Ray, 36, 61, 262n4(ch4)

Rhea, 42f

RNA (ribonucleic acid): alternative biochemistries, 14, 216–19; breakdown of, 14; as information molecule, 142, 216–17; pre-RNA and development of, 142–43, 147; RNA-world scenario, 142–43, 148, 217, 218

robotic spacecraft, 12–13, 254, 258–59

robust biosignatures, 236–37, 242

rocket equation, tyranny of, 249–50

rogue planets, 133–34

Roosevelt, Teddy, 253

Roth, Lorenz, 176

rotorcraft, 256

Russell, Mike, 157, 159

Sagan, Carl, 226, 243, 244

salts: Enceladus, 101–2; Europa, 65–68, 173, 179; freezing point suppressed by, 173; Titan, 119

Saturn: *Cassini* mission, 97; magnetic field, 97, 118; moons orbiting, 13; rings of, 101, 106; seasons, 114f

Saturn moons. *See* Enceladus; Mimas; Titan

S-band, 74

Schumann resonance, 118–19, 263n4(ch7)

Schwinger, Julian, 81

seasons, on Titan, 113–15, 114f

semiconductors, 203–4

sensory systems, 186–92

serpentinization, 154–59, 197

Shirley, Jim, 67

Shock, Everett, 146

shrimp swarms at hydrothermal vents, 196–97, 265n10

sight, evolution of, 186–89

silica, 52, 102, 215, 266n14

silicates: Callisto, 127; as class of minerals, 214; on Earth ocean floor, 91; Ganymede, 123, 127; geology vs. biology and, 213

silicon: building block possibility, 213–15; carbon–silicon bonds, 216; shells of diatoms, 215, 266n14

silicon dioxide, 214–15

small organics, 100

Snake Pit hydrothermal vents, 196, 197

snow, 177

snowflakes, 31, 32f

snow lines, 40–41, 41f, 44

sodium chloride, 68

solid-state greenhouse effect, 129

solvents, 115, 209–12

Sotin, Christophe, 124

sound waves, 58

Southwest Research Institute, 98, 100

Sparks, Bill, 176

specificity of organics, 239

spectroscopy: about, 50–51, 57–58; chemistry connected to (Bunsen, Kirchhoff, and Herschel), 56–59; of Earth (*Galileo*), 243–44; Europa *Clipper*, 252; Fraunhofer's 6-lamp spectroscope, 51–56; Gas Chromatograph–Mass Spectrometer (GCMS), 233–35; infrared spectrum of Europa, 59–65; Ion and Neutral Mass Spectrometer (INMS), 98–104; non-ice material (salts or sulfuric acid) on Europa, 65–68

Spencer, John, 167–68

Spirit, 14

Sputnik Planitia, Pluto, 131–32

squibs, 65

Squyres, Steve, 262n4(ch4)

Stevenson, David, 133, 172

Stofan, Ellen, 256

Stone Age, 199

Strange, Nathan, 98

sublimation curves, 41f

subsumption, 179

sulfate: biodomes scenario and, 205; in early Earth atmosphere, 161; Epsom salts, 66; on Europa surface, 66, 67; ice cycling on Europa and, 180; as oxidant, 146, 167; photosynthesis and, 164

sulfide: in early solar system, 40; hydrothermal vents and, 148; as oxidant, 146; snow line for, 40, 41f

sulfonamide, 221

sulfur: Callisto, 127; Europa, 43; Ganymede, 123

sulfuric acid, 66–67, 209–10

Sun: elements in, 57; magnetic field of, 165; photosynthesis and, 162; spectrum of, 54–56, 55f

technology. *See* intelligence and technology, evolution of

technomimicry, 201–5

Tee-Van, John, 20

Tethys, 42f

tetrahedral pyramid, 29–30, 30f

theodolites, 53

Thomas, Peter, 104, 105

Thompson, Reid, 243

Thomson, Charles Wyville, 19

tidal bulge, 34–35, 35f, 105

tidal heating: Europa, 35, 36, 174–76, 264n4(ch11); Ganymede, 90; in new Goldilocks model, 32–38; Triton, 130

tidepool/warm pond scenario, 138, 148–50

time: Enceladus and question of, 106–7; as origin-of-life parameter, 139–40

tin, 215

Titan: atmosphere, 109–12, 119; future landing missions, 255–57; hydrothermal vents scenario and, 150; ice-covered oceans on, 10, 13; ice shell thickness, 117, 263n4(ch7); k_2 Love number and gravity signature, 116–17; landscape of, 108; life, likelihood of, 108, 119; lightning hypothesis, 118; methane meteorological cycle, 108–9, 113; in new Goldilocks zone, 45; nitrogen and carbon, 44; ocean beneath surface, evidence of, 115–19; ocean depth, 117; Schumann resonance, 118–19, 263n4(ch7); seasonality, 113–15, 114f; simulation of rocks from Earth, 18; size and density, 42f, 44–45; surface images, 111f; temperature and pressure, 108; "weird life" in hydrocarbon lakes, possibility of, 113–15, 119, 210

Titan Mare Explorer (TiME) (proposed), 256

Titania, 13, 42f, 120

Tombaugh, Clyde, 131

top-down approach to origin of life, 143, 146, 147–48

tow-yo-ing, 189–90

transistors, 202–3

Trieste, 21–22, 247, 261n6

Triton: cantaloupe terrain, 128; density and radiogenic decay, 130; evidence for ocean, 128–31; ice-covered oceans on, 10, 13; plume material eruptions from, 128–30; retrograde orbit and tilt, 129; size and density, 42f; surface age and composition, 130–31; tidal heating, 130

Trumbo, Samantha, 67–68

Trump, Donald, 268n10

Tsiolkovsky, Konstantin, 268n3

Turtle, Elizabeth (Zibby), 256

2001: A Space Odyssey (book and film), 262n3(ch4)

2010: Odyssey Two (book and film), 262n3(ch4)

ultra-stable oscillator (USO), 74

Umbriel, 42f

University of Stuttgart, 98, 100

Uranus system, 13, 42f, 120, 257

Urey, Harold, 143

Vance, Steve, 124

Venus: dry and rocky, 40; *Galileo* slingshot around, 72; Goldilocks model and, 25–26; loss of water, 37; *Mariner 2* flyby, 12, 21; origin of life conditions and, 150; robotic exploration of, 208; sulfuric acid on, 210

Viking landers (Mars), 231–37

volatile compounds, 40–41

volcanic activity: cryovolcanism, 110, 130, 131; Ganymede, 122–23; Io, 36, 61

Voyager 1 and 2: Enceladus images, 96–97; Europa images, 23, 65–66; Io images, 36; Triton images, 128; water likelihood and, 13

Walsh, Don, 21–22

warm pond/tidepool scenario, 138, 148–50

water: atomic structure and covalent bonding of water molecules, 30f; change in density across the liquid–solid boundary, 28; electromagnetic spectrum and, 59; as essential to life, 12; ice, floating and crystal structure of, 25–32; likelihood of, on moons, 13; loss of, on Venus and Io, 37; in past vs. present, 13–14; polarization of, 115; as solvent, 209; water–rock interactions, 43, 124, 130. *See also* ice; oceans of the outer solar system

"weird life," 18, 113–15, 119, 210

Westheimer, Frank, 223

Whitesides, George, 4–5

wobbles, 104–6

Woese, Carl, 232